THE

Mosquito
BOOK

An Entertaining, Fact-filled Look at the Dreaded Pesky Bloodsuckers

By
Brett Ortler

Adventure Publications, Inc.
Cambridge, MN

Dedication

For my parents, and my sister, Emily

Acknowledgments

I'm grateful for the Walter Reed Biosystematics Unit and their vast array of mosquito-related literature. *PLOS ONE* and *Malaria Journal* are very useful sources of information, as is the National Institutes of Health's PubMed database. The CDC's website is also a wonderful resource, as is the CDC's Public Health Image Library.

Special thanks to Sandy Brogren, chief entomologist at the Metropolitan Mosquito Control District, for reviewing the book for accuracy.

Book and cover design by Lora Westberg

Photo credits are listed on page 142.

10 9 8 7 6 5 4 3 2
Copyright 2014 by Brett Ortler
Published by Adventure Publications, Inc.
820 Cleveland Street South
Cambridge, MN 55008
1-800-678-7006
www.adventurepublications.net
ISBN: 978-1-59193-488-2

Disclaimer: When using insecticides or repellents, it is your responsibility to always read and follow the instructions on the label, and to use the products in an appropriate manner. Always consult with a doctor prior to selecting an insecticide or repellent or if you have questions about how to use a product or the correct concentration to choose. **This book is not intended as a substitute for professional medical advice.**

If you suspect someone has inadvertently absorbed, ingested or inhaled an unhealthy amount of any insecticide or repellent, call 911 and your local Poison Control Center. To reach one, dial: 1-800-222-1222.

Table of Contents

Frequently Asked Questions 8

Introduction 10

The Basics 12

　What Are Mosquitoes? 14

　Mosquito Anatomy 16

　Total Number of Mosquito Species Worldwide 17

　Number of Species in the Continental United States 18

　The Mosquitoes to Worry About: The "Flying Syringes" 20

　What's in a Name? 22

　Even Mosquito Researchers Get Annoyed at Mosquitoes 23

　Four Phases of Life 24

　Where Do Mosquitoes Lay Eggs? Pretty Much Everywhere 26

　Floodwater Mosquitoes 26

　Freshwater and Stagnant Water Mosquitoes 27

　Eggs: Laid Alone or Together? 29

　Waiting It Out: Dormancy 30

　The Eggs Hatch and Larvae Emerge 31

　Let's Go Snorkeling! 32

　Pupation: Real-Life Transformers 33

　Surprise, Adult Mosquitoes Are Vegetarians! 34

　Mosquito Mating: Music (and Dancing) in the Swarm 36

　Hibernation/Overwintering 37

Honing In 39

　Heat, Humidity, Movement and Dark Clothing 40

　On the Clock 41

It's Not Paranoia If Bloodsucking Insects
Are Really After You. 42

A Mix of Microbes 43

The Problem of Beer 44

Mosquito Immunity and Natural Repellents 45

"Stress" May Be Best 46

The Perfect Swarm 47

Warp Speed, Mr. Sulu! 48

Target Acquired: Pregnant Women 49

The Bite: What Happens When a Mosquito Finds You 51

How They Bite 53

Mosquitoes Have a Big Pouty Lip. Really! 54

The Bite Itself is Actually Painless 55

Mosquitoes, *Star Trek* and Microneedles 56

Pumping Blood. 57

Eating Overload 58

Population Density: How Many Mosquitoes Are There on
an Average Night? 59

The Worst Party Ever: An (Involuntary) Blood Drive 60

Exsanguination: The Worst-Case Scenario 61

Pity the Caribou 62

Watch Out for Grandma Skeeter! 63

Once Bitten. 64

Why Some Years Are Worse Than Others 65

The Genetics of Vulnerability to Mosquitoes. 66

Mosquitoes as Disease Vectors: Flying Syringes 69

Malaria: A Deadly Package Delivered by a Mosquito 70

Attack of the Clones 72

A Terrible Toll. 73

Humanity and Malaria: Familiar Foes 74

Evolution in Action: The Malaria Hypothesis 75

Quinine and Sweet Wormwood Save the Day 76

Efforts to Eradicate Malaria. 77

Yellow Fever and Dengue Fever 78

West Nile Virus: New Kid on the Block. 79

Mosquitoes and Encephalitis 80

Heartworm. 81

Avoiding Mosquitoes Made Easy. 83

Get Rid of That Standing Water! 84

Location, Location, Location—and Time 86

What's a Mosquito's Favorite Color? 87

DEET: The Gold Standard 88

General Tips for Using DEET 89

Pyrethrum, Pyrethrins and Pyrethroids:
  Chrysanthemums to the Rescue. 90

Permethrin is for Use on Clothing, Not Skin. 92

General Tips for Using Permethrin on Clothing 93

Using Other Types of Permethrin Insecticides. 95

Picaridin: Another Product Inspired by a Natural Repellent . . . 96

General Tips for Using Picaridin. 97

Eucalyptus Isn't Just for Koalas: Oil of Lemon Eucalyptus,
  a Natural Mosquito Repellent 98

General Tips for Using PMD and Oil of Lemon Eucalyptus 99

IR3535: A Clunky Name, but a Popular
Choice Across the Pond. 100

General Tips for Using IR3535 101

**What Doesn't Work: Superstitions,
Mosquito Traps and Other Bogus "Cures"** 102

Mosquito Control on the Home Front:
Mosquito Dunks and Citronella. 108

**A Mosquito Miscellany: The Weird,
the True, and the Funny of Mosquito Lore** 110

Mosquitoes in War 112

Bugs in Battle: Entomological Warfare! 113

Time for Payback:
The World Mosquito Killing Championship 114

Annoyance, Meet Terror: The Botfly 115

An Ongoing Science 116

New Approaches and Treatments 117

Invasive Species and Mosquitoes in a Warming World 118

Getting Involved. 119

Identifying *Aedes*, *Anopheles* and *Culex* Mosquitoes 120

How to Identify Potential Disease-Carrying Mosquitoes 122

Identifying Eggs 122

Identifying Larvae 123

Identifying Adults 124

Table of Repellents/Insecticides 126

Recommended Resources/Reading. 128

Bibliography 129

About the Author 142

Frequently Asked Questions

Why do mosquitoes bite?

Mosquitoes don't actually obtain nutrition from blood. Instead, female mosquitoes use the blood to obtain the proteins needed to produce a batch of eggs. Without a "blood meal" most species can't produce eggs.

How many mosquitoes are there in my yard?

This is one of the most common questions about mosquitoes, but it is also one of the hardest to answer. The reason is pretty obvious: mosquitoes are small, hard to track, and populations vary greatly by species, habitat, and light and weather conditions. This is all too easy to observe—your front yard might be miraculously mosquito-free, but mosquitoes might chase you out of your garden. Suffice it to say: we're seriously, seriously outnumbered.

What's the most effective repellent?

Not all mosquito repellents are created equal. DEET is considered the "gold standard" of repellents, so much so that all others are compared to it. A number of other effective repellents are now on the market, including Picaridin, Oil of Lemon Eucalyptus (aka PMD) and IR3535. Take note, however, that not every repellent works equally well against every species of mosquito, and they vary wildly by concentration and how long they protect you from mosquitoes. What's more, some should not be used on young children. Always read the product labels and heed the directions when choosing a product. **When in doubt, consult your doctor.**

How can I fight back against mosquitoes?

Know your enemy. If you know where mosquitoes develop (standing water!), you can give them fewer places to develop. If you know when they are most active (dawn and dusk), you can avoid them at their worst. And if you know what to wear and which repellents to use, you can help protect yourself from being bitten.

What diseases can mosquitoes transmit?
Quite a few of them, unfortunately. In the tropics, mosquito-borne diseases represent one of the greatest health threats, as mosquitoes transmit malaria, yellow fever, Dengue fever, and a host of other diseases. In the U.S., we're mostly spared from the worst of these diseases, though travelers can contract them overseas, and outbreaks of some (Dengue, for instance) are not unheard of, especially in areas such as Florida and Hawaii.

Unfortunately, a number of other mosquito-borne diseases are widespread in the U.S., including West Nile virus and a number of varieties of encephalitis. Outbreaks of these diseases generally occur in the summer (when mosquitoes are more active), and there are thousands of cases each year, leading to many hospitalizations and even some deaths.

If I trap a mosquito's proboscis and/or flex a muscle in the vicinity of the area where a mosquito lands, can I cause it to ingest too much blood and explode?
Nope. When a mosquito is filled with blood, its nervous system essentially has a "shut-off valve" that tells it to stop ingesting more blood. So if you hold the mosquito there or flex the muscle, it won't ingest more blood. However, in what must have been a satisfying experiment, researchers have successfully severed that nerve connection, which led mosquitoes to ingest several times their capacity and eventually explode.

Can mosquitoes transmit HIV?
Thankfully, no. The HIV virus cannot survive[1] in a mosquito; plus, they only ingest a very small amount of the virus.

Introduction

Mosquitoes may be the most loathed creatures in the insect world
—and for good reason. They can easily make a barbecue miserable
or ruin a summertime stroll. Worse yet, they transmit diseases
that kill millions of people each year and cost the world economy
billions of dollars.

Nonetheless, they are also a resilient, wildly diverse group of
creatures, and if you take a closer look, mosquitoes are really
fascinating creatures. Oh, what am I saying? They are absolutely

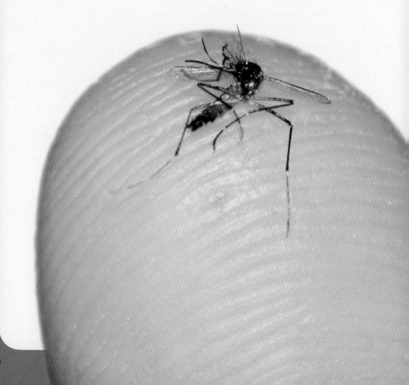

annoying! If the opportunity presents itself, I wholly encourage you to use this book to take a few out.

That's also the reason I wrote this book; more than anything, I want to help you—the everyday reader—avoid mosquitoes. The best way to do that is to learn about mosquitoes; that's why this book covers everything from the basics of the mosquito's life cycle and how mosquitoes seek out hosts, to the "Big Three" medically important types of mosquitoes and the repellents that can help keep them away. I'm also including a sampling of fun mosquito-related facts and a collection of mosquito-related superstitions that never seem to go away. (Hint: Listerine won't get rid of mosquitoes.)

In short, I hope this book helps you laugh a little and learn a lot, but most importantly, I hope it helps you escape the worst that mosquito season has to offer.

The Basics

If you want to know how to avoid mosquitoes, you need to learn about the different types of mosquitoes found in North America, their life cycle, and perhaps most importantly, the basics about mosquito habitat and behavior. Consider this section of the book your primer to the world of the mosquito.

What Are Mosquitoes?

Mosquitoes are insects. Specifically, they are small flies in the scientific family Culicidae.

For those of you needing (a brief!) refresher on your biology: Scientists use the taxonomic system to classify and catalog life. There are eight categories, or ranks, in it. The general idea is pretty simple: the lower the rank, the more specific you get.

As you can see, just a quick look at the taxonomic system tells us a lot about mosquitoes. We can immediately see that they are insects, and more than that, they are a specific type of insect—true flies—which means that mosquitoes belong to the same order as black flies and the common housefly.

Domain—Eukaryota

   Kingdom—Animalia

      Phylum—Arthropoda

         Class—Insecta

            Order—Diptera
              (true flies, including mosquitoes)

              Family—Culicidae (mosquitoes)

                Genus

                  Species

Mosquito Anatomy

All insects, including mosquitoes, have three main body parts: a head, a thorax and an abdomen. They also have three sets of legs and one or two pairs of wings.

Like other flies, mosquitoes have a number of features[2] in common: they have one pair of wings, large compound eyes, mouthparts adapted for sucking (or in the case of mosquitoes, piercing), a pair of simple antennae (though mosquito antennae are often long and frilly) and a pair of (often tiny) halteres, small stump-like wing remnants that act as flight stabilizers.

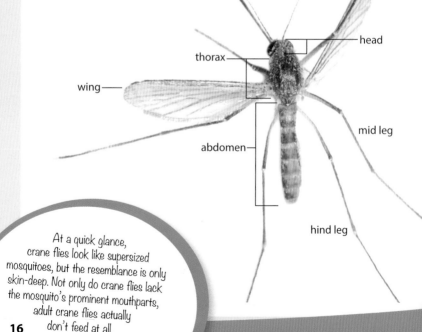

foreleg

head

thorax

wing

mid leg

abdomen

hind leg

At a quick glance, crane flies look like supersized mosquitoes, but the resemblance is only skin-deep. Not only do crane flies lack the mosquito's prominent mouthparts, adult crane flies actually don't feed at all.

Total Number of Mosquito Species Worldwide

There are over 3,500 mosquito species[3] worldwide. About half of all species are endemic, which means that they are only found in one country or area. The other half is more widespread, with some species found across thousands of miles.

As mosquito populations are dependent upon heat and the presence of water, mosquitoes are dependent on geography. Generally speaking, the closer one gets to the equator, the more mosquito species you'll find[4]. (Countries with large areas also tend to have more mosquito species.)

For example, Norway is home to 16 mosquito species. Brazil has over 450.

While we generally think of mosquitoes as drab pests, some mosquitoes are actually gorgeous. Perhaps the most famous example is the species *Sabethes cyaneus*, which is renowned for its iridescent coloration and its feather-like decorations. If you were bitten by a mosquito like that, at least you'd get a good show!

Number of Species in the Continental United States

The Continental U.S. is home to 166 mosquito species[5], but the number of species per state and mosquito population density vary widely. Southern states, such as Texas (85) and Florida (80), often have more mosquito species, but not by all that much. Northern states, such as Minnesota and New Jersey, have quite a few as well, 51 and 63, respectively. Even comparatively warm, dry states, such as Arizona and Nevada, have robust populations that can wreak havoc. In short, there is almost no escaping them, so you need to know how to fight back.

Of course, Brazil's many mosquito species won't necessarily ruin your vacation in Rio. After all, species numbers don't say much about population numbers. A case in point: Alaska and Minnesota are home to far fewer species, but they definitely boast more than their fair share of individual critters in "skeeter season." Unsurprisingly, both Alaska and Minnesota have mosquitoes named after them.

Culex

Anopheles

The Mosquitoes to Worry About: The "Flying Syringes"

All mosquitoes belong to one large family of insects—Culicidae. This family consists of two large groups, which are called subfamilies: the Anophelinae (480 species) and the Culicinae (3,000 species). Some species don't bite humans at all, but a good number are serious pests and a few troublemakers are responsible for spreading serious diseases.

In particular, three groups of mosquitoes are responsible for the vast majority of mosquito-borne diseases. These bad apples are often called the "Flying Syringes."

Because of their medical importance, it's helpful to be familiar with them. As it happens, these three groups also include some of the primary pest species—so even if they don't make you sick, they might still ruin your barbecue.

Aedes

Anopheles mosquitoes are members of the Anophelinae subfamily. One genus of the Anophelinae mosquitoes—the genus *Anopheles* —is notorious for spreading malaria. In all, there are 12 species of *Anopheles* in North America[7].

Both the *Aedes* genus and the *Culex* genus are members of the Culicinae subfamily. There are 11 species of medically important *Aedes* in North America[8], and 9 species of medically important *Culex*.

The *Aedes* mosquitoes spread yellow fever and Dengue fever.

Culex mosquitoes spread West Nile, a number of types of encephalitis[9] and several parasitic diseases.

The good news? Cases of malaria, yellow fever and Dengue fever are mercifully rare in the U.S., and those infected have usually traveled abroad. Other diseases—including West Nile—still remain a serious concern, however.

What's in a Name?

The word *Anopheles* (a-NOF-o-leez) comes from the Greek word *anōphelēs*, which translates to "useless.[10]"

The word *Aedes* (AY-deez) also comes from Greek, from the word *aēdēs*, which translates to "unpleasant.[11]"

The genus name *Culex* (CUE-lex) is far less colorful and simply means "gnat" or "fly" in Latin[12].

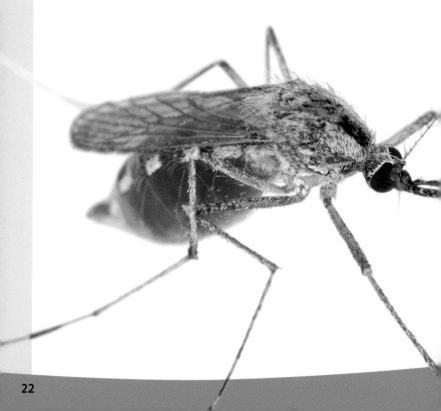

Even Mosquito Researchers Get Annoyed at Mosquitoes

Some mosquito researchers clearly had some pointed (ha?) opinions about mosquitoes. Want proof? The species names they chose say it all; you'll probably recognize most of these root words, all of which are used in English: *vexans*[14] and *inundatus* and *abominator* and *excrucians* and *tormentor* and *horrida*. None of them have positive connotations.

Some biologists literally agree with *Anopheles'* scientific name and think that mosquitoes are worthless, even calling for the outright eradication of mosquitoes. They argue that the ecosystem would likely be relatively unaffected[13].

Four Phases of Life

Like butterflies, ants and many other insects, mosquitoes go through four general stages of life: they start life as an **egg**, which hatches into a **larva**. After feeding, the larva **pupates**, where it transforms into an **adult**.

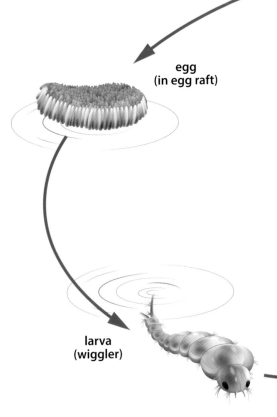

**egg
(in egg raft)**

**larva
(wiggler)**

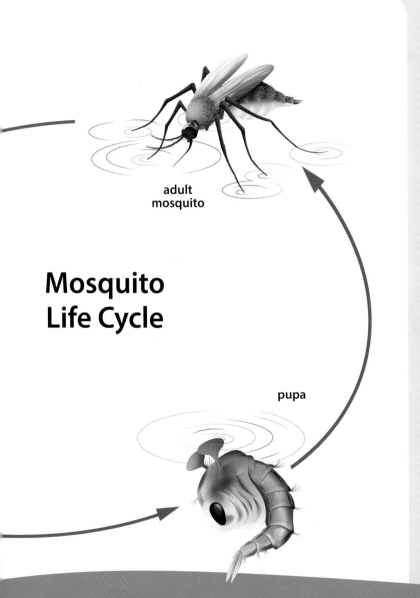

adult
mosquito

Mosquito
Life Cycle

pupa

Where Do Mosquitoes Lay Eggs? Pretty Much Everywhere

Female mosquitoes lay their eggs on or near water, or in areas that will soon be exposed to water. As you might expect, that encompasses much of the U.S., and with over 150 species in the country, there's a great deal of variation among mosquitoes.

Very generally speaking, there are two types of mosquitoes—floodwater mosquitoes and freshwater/stagnant water mosquitoes. These names are informal, and since mosquitoes are quite opportunistic when depositing eggs, some species undoubtedly fit in both categories. Still, these categories are enough to give the layperson an idea of where mosquitoes lay their eggs—and how one can avoid giving mosquitoes places to deposit eggs.

Floodwater mosquitoes

As their name suggests, floodwater mosquitoes (which includes important *Aedes* species) lay their eggs in areas that are damp or that will be inundated by water, such as salt marshes or riverside forests. The eggs need to be inundated by water before they hatch, but reduction of dissolved oxygen in the water is the most important factor for egg hatching[15]. When levels drop beyond a certain point, the eggs hatch. This ensures that the larvae will be born into the shallow, food-filled areas where the larvae thrive.

These mosquitoes often become a scourge after river flooding or the storm surge of a hurricane.

Freshwater and Stagnant Water Mosquitoes

Some species, like the malaria-bearing *Anopheles* mosquitoes, only lay eggs on freshwater, whereas the notorious pest *Culex pipiens* lays eggs[16] in stagnant water, everywhere from cesspits and plant pots to hoofprints in cow pastures. The invasive Asian tiger mosquito will deposit eggs nearly anywhere it can —and doesn't venture far from where it hatches, so if a female of a pest species deposits eggs in the birdbath on your deck rail, you're certainly on the dinner menu[17]. Perhaps surprisingly, flowing streams and deep ponds and lakes are often less ideal environments for most mosquito species, as they have less food, too much turbulence and more predators.

To make matters more complex, some mosquitoes are often referred to as "container" mosquitoes because they are found in a variety of natural or artificial containers. Mosquitoes

How opportunistic are mosquitoes when finding a place to lay their eggs? During a study in the 1930s, a cage holding research mosquitoes and eggs was inadvertently torn open. When the researchers returned and discovered the damage, they noticed that some mosquitoes had escaped, but wild females had entered the cage and laid additional eggs[18] !

found in natural containers are often referred to as "woodland" or "treehole" mosquitoes because they are found in wooded areas, and often deposit eggs near trees—or in cavities within them. When it rains, these holes fill in with water, and the eggs then hatch. There are many types of natural cavities—from rotting stumps to containers as specific as a particular type of pitcher plant. In fact, some mosquito species even lay their eggs in pitcher plants, which serve as natural cavities and catch rainwater[19]. An African *Aedes* species even lays its eggs on the legs of crabs, which then transport the eggs to the crab's nest, where the eggs later hatch[20].

Of course, many[21] mosquitoes are just as comfortable depositing eggs in artificial containers such as gutters, tarps, birdbaths and old tires. Oddly enough, many disease-transmitting mosquito species have been found developing in artificial containers in cemeteries. The vases, toys, and flowerpots left behind by mourners prove to be ideal habitat for these sometimes-deadly insects[22].

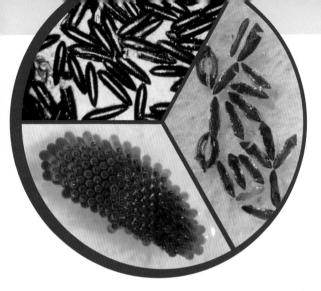

Eggs: Laid Alone or Together?

In North America, there are two primary ways that mosquitoes lay eggs: some mosquitoes, including all of the *Anopheles* and *Aedes* species, deposit their eggs one-by-one, either on the water or on moist soil at the water's edge, whereas nearly all[23] *Culex* species (and those of a few other groups) glue their eggs together to form an egg raft, which floats on the water and rebounds to the surface when submerged.

When female mosquitoes build an egg raft, they are nearly totally committed. According to the observations of one researcher, *Culex* females were so focused on building an egg raft that one could be picked up from the surface and didn't try to fly away[24]. Talk about a maternal instinct!

Waiting It Out: Dormancy

If the conditions aren't right, the eggs of some species can become dormant; dormancy can last for some time—the eggs of some *Aedes* species[25] can remain viable for up to four years. When flood-water eggs are laid in cooler weather, the eggs remain dormant until the weather warms. This is also why Dengue fever and other diseases carried by *Aedes* mosquitoes often spike in spring or during the monsoon season in tropical areas.

Whereas most mosquito females simply deposit their eggs and leave, one species in South America actually broods its eggs, protecting them until they hatch[26]. While this is common in other insects, it's the only mosquito species known to do so.

The Eggs Hatch and Larvae Emerge

Mosquito eggs usually hatch within a few days, but under ideal[27] conditions, some species can hatch within just 30 hours. Larvae are colloquially referred to as "wigglers" thanks to their wiggly swimming motions. Larvae have two primary jobs: eating and molting. Larvae go through four molting periods, each of which is referred to as an "instar." After hatching, the larva is in its first instar. Once it molts, it is in the second instar, and so on.

With each molting period, they get larger. Of course, to grow, they need food, and eating is their specialty. Larvae are almost always filter feeders and eat a variety of algae, single-cell organisms and plant material. In some *Culex* species, larvae are fierce predators, consuming many larvae of other mosquito species, and even members of their own species[28].

The length of the mosquito's larval stage depends on the species and the conditions but is usually a matter of 7–10 days[29].

Let's Go Snorkeling!

All mosquito larvae are aquatic, but they still need to breathe air. So how do they do it? *Aedes* and *Culex* mosquitoes essentially use a snorkel, which is technically called a siphon. When resting, they hang down from the snorkel. *Anopheles* species lack this snorkel and instead lie parallel against the water's surface, breathing through tubes on their abdomen[30].

Some larvae have even more ingenious methods of breathing. Mosquitoes in the *Mansonia* genus are wholly aquatic. They drill into underwater plants[31] in order to obtain oxygen.

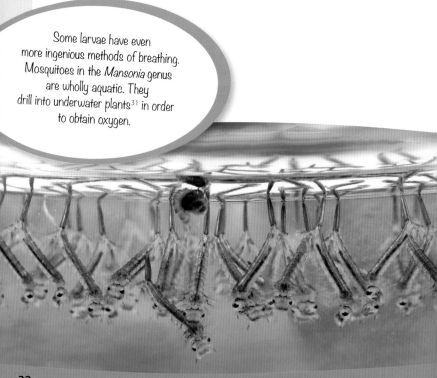

Pupation: Real-Life Transformers

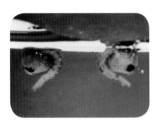

Like moths and butterflies, mosquitoes pupate, but mosquitoes pupate in water. There they develop into adults and then somehow manage to emerge from the water's surface and fly away. The transformation is like something out of science fiction; when they go in, they are worm-like aquatic creatures that get around primarily by wriggling and that eat phytoplankton and one-celled organisms. When they emerge as adults, they are six-legged flies that subsist on an entirely different diet—sugars from plants.

Pupation periods vary significantly by species and the local conditions—generally, warmer conditions speed things up, but adults usually emerge in a matter of days.

The pupae of one type of mosquito species are reported to glow in the dark. Found in Brazil, these species glow slightly in a purplish hue[32].

Surprise, Adult Mosquitoes Are Vegetarians!

Once the adult mosquitoes emerge, they have two main jobs: eating and mating. Mating usually happens first and takes place at or near the site of emergence. After mating, the females forage for a blood meal; they do so because the proteins found in animal blood—specifically chains of amino acids—help them[33] to produce more eggs. Without a blood meal, most species can't produce eggs. There are a few exceptions; such species are called autogenous and can produce eggs without a blood meal[34].

When most female mosquitoes seek out a blood meal, humans often aren't their preferred species. When given the choice, mosquitoes often prefer livestock, such as sheep[36]. So why do they pester you at your barbecue? Because we are often the most numerous host in their habitat.

So what do mosquitoes actually eat? Sugars, usually from nectar or fruits. In the lab, scientists often feed adult mosquitoes a simple mixture of sugar water[35], though research mosquitoes have also dined on corn syrup, fruit juices and raisins, among other things.

Mosquito Mating: Music (and Dancing) in the Swarm

Mosquitoes usually mate in swarms—large collections of males that are frequented by females[37]. As it turns out, it's a surprisingly musical affair. The female mosquito's characteristic buzzing sound is actually music to the male's ears, and they are attracted to the sound. (In fact, in 1948, researchers armed only with a tuning fork tuned to the female mosquito's frequency were able to attract male mosquitoes.)

The males in the swarm are pretty musical as well—as they "dance"[38] amid the swarm, and when the male and female approach one another, they harmonize, creating the equivalent of a mosquito love song[39]!

The *Anopheles* mosquitoes don't mate right away because they have to undergo some ... changes[40]. In what must be a terrifying process, during the first 12-24 hours of the male's adulthood, his genital organs are inverted—turning a full 180 degrees. This orients them correctly so mating can occur.

Hibernation/Overwintering

So how do mosquitoes survive the winter? It depends. Some species overwinter as adults, usually in a well-sheltered area, such as a cave or a culvert. Other species survive in the egg stage or as larvae. Either way, the mosquitoes are raring to go when the weather warms up.

St. Paul, Minnesota, has many well-known caves. A mosquito researcher happened upon one of these caves, which was being used to grow mushrooms, and discovered millions of hibernating[41] adult mosquitoes.

Honing In

How do mosquitoes find their targets in the first place[42]? The answer is much more complicated that one might think: mosquitoes find their hosts thanks to a variety of environmental cues. They are adept at detecting carbon dioxide and heat, and they have a complex sense of smell that helps them sense the many chemical compounds that potential hosts emit. Mosquitoes can also sense movement from a distance and often seek out hosts in twilight or under the cover of darkness. (Some species do bite in full daylight, however.)

Heat, Humidity, Movement and Dark Clothing

In one sense, mosquitoes aren't all that different from heat-seeking missiles, as both hone in on heat. A mosquito's antennae contain heat receptors, which pick up body heat, and this is one way they find a potential host. This is also why mosquitoes are drawn to people wearing darker colors[43], as darker colors retain more heat, giving the wearer an even bigger heat signature. (Lighter colors avoid this problem—and that's why loose, baggy light-colored clothing is recommended in peak mosquito season.)

According to a study[44] of *Aedes* mosquitoes from 1938, the order of color preference—from most attractive to least attractive—was black, red, grey and blue, khaki, green, light khaki and yellow.

Movement[45] and humidity seem to play a key role in attracting, too. In a number of studies, moving targets—or those overlain with moving patterns[46]—were more likely to entice female mosquitoes to land. In a study involving robotic targets and *Aedes* mosquitoes, they landed more often on robots that were both humid and warm.

Why don't mosquitoes like the cold? It prevents them from flying. Mosquitoes also happen to be cold-blooded. Thanks to their role in spreading deadly diseases, this literally makes them cold-blooded killers.

On the Clock

Anyone who's had an evening in the backyard ruined by mosquitoes knows that there are specific times when mosquitoes are most common. Generally speaking, mosquitoes tend to be most active in the early morning and in the early evening, though some species bite all day long. As it turns out, mosquito behavior patterns are hardwired into their biology; they contain what is called a "clockwork gene"[47] that makes them "preset" to a 24-hour schedule (technically known as a Circadian rhythm). Mosquito activity is particularly tied to evening, especially sundown. It's probably no accident that this time frame coincides with when their primary hosts—birds and dogs and you—are active.

It's Not Paranoia . . .

. . . If Bloodsucking Insects Are Really After You

So what happens when a mosquito finds a group of humans? Does it select a host at random? As it turns out, the answer is no; mosquitoes are often picky about choosing a host, and research indicates[48] that they choose their specific hosts primarily by scent, honing in on specific olfactory cues. Human skin and breath emit several hundred chemical compounds, and as gross as it may sound, each of us emits an "odor plume." Mosquitoes follow this plume back to choose a host[49].

So if you ever thought you were being singled out by mosquitoes, maybe you weren't being paranoid. Mosquitoes do really prefer some people.

A Mix of Microbes

There are a number of factors that affect which hosts mosquitoes choose, but specific types of bacteria seem to play a role. Human skin is positively teeming with bacteria—the average person's skin is home to perhaps 1 trillion bacteria[50]—and those bacteria affect how we smell. Want proof? Forgo the deodorant for a day; that less-than-pleasant scent is caused by bacteria metabolizing your body's sweat. (Without the bacterial input, sweat is actually odorless.) According to recent research[51], mosquitoes are more attracted to people with specific types of bacteria. For people in areas ravaged by malaria, the right bacterial "defense system" might even help people avoid infection, as they'd be bitten less often and therefore be less likely to contract mosquito-borne disease.

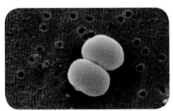

Staphylococcus epidermidis, one of the most common bacterial species on human skin

Even though fetuses are sterile, human infants develop a stable bacterial community within[52] 48 hours. When you combine the number of bacteria inside a human body and on its surface, there are more bacterial cells than human cells. That's not necessarily a bad thing, as our microbiome plays an important role in all sorts of functions (including in our gastrointestinal system). Without our bacterial friends, we'd certainly all be dead.

The Problem of Beer

Human skin and breath emit several hundred chemical compounds, and that means there are many possible factors that could attract mosquitoes. To make matters more complicated, not everyone emits the same ones, as they can vary for a number of reasons, including one's genes, lifestyle and even one's choice of beverages. In particular, scientists have singled out one drink that has been shown to attract malaria-carrying mosquitoes: beer. In a study performed in Burkina Faso, Africa, mosquitoes landed more often on study participants who drank beer than on those who didn't[53]. As it turns out, this has serious consequences, as beer consumption in malaria-stricken areas is increasing, potentially leading to an uptick in malaria rates.

Mosquito Immunity and Natural Repellents

Just as mosquitoes are attracted to some people more than others, mosquitoes find others far less inviting. Why? After all, everyone produces some compounds that mosquitoes like—carbon dioxide, among others. As it turns out, some of us also produce our own natural repellents, too[54]. These effectively[55] mask the more attractive scents, so the mosquito seeks its blood meal elsewhere.

These compounds have now been isolated and tested and are showing great promise as natural insect repellents.

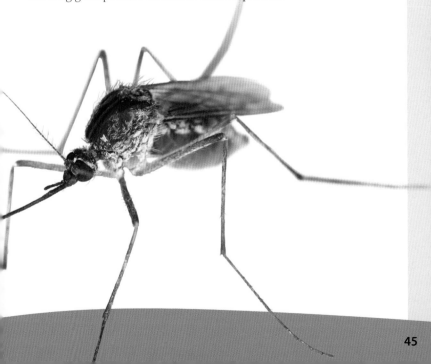

"Stress" May Be Best

A number of the compounds that chase off mosquitoes[56] are emitted when an animal is in stress. So are people who are immune to mosquitoes simply stressed out? Not necessarily— they may simply have a higher baseline than people more susceptible to mosquito bites.

The Perfect Swarm

One way to remember how to avoid mosquitoes is to envision the best possible conditions for mosquito activity—or as I call it, the worst possible time for an evening run. It'd have to take place on a warm, wind-free summer evening, an hour or two before sundown. I picture our poor runner in a tight-fitting black shirt and shorts (to trap heat), and running fairly quickly—to generate ample amounts of heat, carbon dioxide, sweat and lactic acid (mosquito attractants). Then, the unthinkable happens: they cramp up near an often-flooded area or one with lots of stagnant water or containers for mosquitoes to develop in (a cemetery, say). Maybe this happens after a few weeks of heavy rain.

Needless to say, cramp or not, the runner would undoubtedly get moving pretty quickly once the swarm sets in.

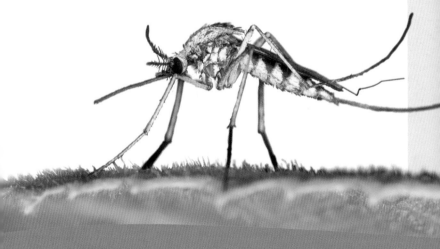

Warp Speed, Mr. Sulu!

Unlike black flies or horseflies, mosquitoes can't fly very quickly. They only reach a top speed[57] of about 2 miles an hour and they sometimes seem to spend as much time hovering as they do flying. With such a low top speed, mosquitoes don't fare well in windy conditions. If the wind is up in your backyard, you might notice that the mosquitoes are less of a bother than usual. They're still present, of course, but they simply can't make headway against the wind.

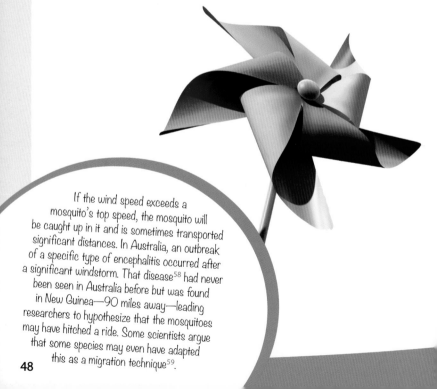

If the wind speed exceeds a mosquito's top speed, the mosquito will be caught up in it and is sometimes transported significant distances. In Australia, an outbreak of a specific type of encephalitis occurred after a significant windstorm. That disease[58] had never been seen in Australia before but was found in New Guinea—90 miles away—leading researchers to hypothesize that the mosquitoes may have hitched a ride. Some scientists argue that some species may even have adapted this as a migration technique[59].

Target Acquired:
Pregnant Women

Malaria-carrying mosquito[60] species specifically seek out pregnant women as hosts. This may occur because pregnant women produce more carbon dioxide than others or because they also emit more warmth, which produces more skin emanations, making them a bigger target. Worse yet, in areas with malaria-carrying mosquitoes, pregnant women often need to leave the protection of their bed nets in order to use the restroom, which as any pregnant woman can tell you, they need to do rather often. This exposes them to more potential mosquito bites, and potentially, to malaria and other diseases.

The Bite: What Happens
When a Mosquito Finds You

As annoying as it is, a mosquito bite is an engineering marvel. It's amazing enough that such tiny insects can locate their hosts based only on a few environmental cues; it's even more amazing that mosquitoes can bite us at all. This section covers the nitty-gritty of mosquito bites: the mechanics of the bite, how much blood they take, why a bite hurts, and just to give you perspective, an example of what a "worst-case" mosquito scenario might look like.

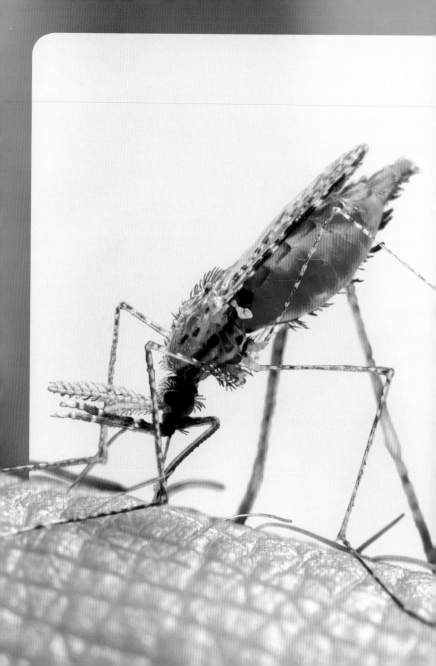

How They Bite

Mosquito bites are annoying, but in a sense, it's sort of amazing they occur at all. After all, human skin is pretty tough stuff. When other animals bite or sting us, they often use some pretty heavy-duty hardware: a rattlesnake's fangs are essentially two hypodermic needles, and a bee or a wasp stinger looks formidable, even from a distance. When acquiring a blood meal, mosquitoes insert a set of "microneedles" that are just a few millimeters long and roughly the diameter of a human hair. The resulting mosquito bite may hurt, but in terms of engineering, it's a pretty impressive feat.

A mosquito's head

Mosquitoes Have a Big Pouty Lip. Really!

Anyone who has caught a mosquito and ripped off its "stinger" knows that mosquitoes acquire blood thanks to the proboscis, which is actually a set of very specialized mouthparts. From our perspective, the proboscis seems relatively simple, but when you look closer—a lot closer, as it turns out—a female mosquito's mouthparts are pretty complex.

When you zoom in, the proboscis consists of a few basic[61] parts: The labium is essentially[62] an extended lower lip—mosquitoes are pouters!—and includes a sheath that covers the complex series of parts that actually pierce the host's skin and draw the blood.

When a mosquito lands, the labium opens to expose the mandibles, which are modified jaws, that are essentially structural supports. When the labium opens, it also reveals the fascicle, a group of very thin stylets, which are essentially tiny needles. These are connected to the labrum, the tube through which the host's blood flows.

The amazing part is that this "Rube Goldberg" device works at all. After all, the "needles" are incredibly fragile. They consist of a small amount of chitin, the same stuff that makes up the exoskeletons of lobsters and other crustaceans.

A male mosquito has similar mouthparts, but its hardware is adapted for ingesting flower nectar and other liquefied sugars.

The Bite Itself is Actually Painless

Here's something that may surprise you: a mosquito's bite is actually painless. Then why do they itch and hurt so much, often seemingly instantaneously? The answer is simple—when a mosquito stings, it introduces a small amount of its own saliva in the process. This saliva acts as an anticoagulant, helping the host's blood flow more freely. Compounds in the saliva cause pain and trigger another aspect of the human body's defense[63] system: the release of histamine, which causes the surrounding tissue to swell up and can cause pain and itching. This means that the pain we feel from a mosquito bite is essentially an allergic reaction.

Mosquitoes, Star Trek and Microneedles

Over the past few decades, we've seen a technology revolution. Devices such as tablet computers and personal computers were once simple props on television shows like *Star Trek*, but today they actually exist, and then some. (An example: a NASA researcher is actually investigating[64] the basic feasibility of a real-life Warp Drive. Really!)

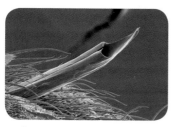

Mosquitoes have led to plans for a similarly futuristic piece of technology—the equivalent of *Star Trek's* "hypospray." As fans of the show know, the hypospray was essentially a painless delivery device for medicine. Thanks to the mosquito's unique mouthparts—which pierce the skin without causing pain—we may be able to replicate this, perhaps doing away with shots altogether.

If the technology pans out, the mosquito, a perpetual cause of pain and misery, could actually lead to pain-free vaccinations at the doctor's office.

Pumping Blood

How do mosquitoes pump blood to begin with? Well, their head contains a pair of pumps that work in tandem to extract blood from the host. These pumps are fast and efficient and can pump up to three times the mosquito's own body mass[65].

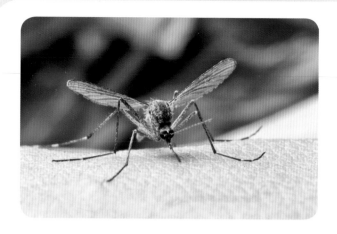

Eating Overload

Sadly, despite many claims to the contrary, it's not possible[66] to cause a mosquito to explode if you trap its proboscis or flex your muscles near the mosquito. In what may have been the most satisfying—albeit gross—study ever, researchers *were* able to get mosquitoes[67] to explode after forcing them to ingest too much blood. To do so, however, they had to sever a nerve in the mosquito's abdomen. When the mosquito can no longer carry any more blood, the nerve is activated, telling the mosquito to stop. When that nerve is cut, they never get the message, and they quite literally "drink" until they explode. Before they do, however, they ingest three to four times more blood than normal.

Population Density

How Many Mosquitoes Are There on an Average Night?

Estimating an area's mosquito population is difficult, as population totals vary wildly. Very generally speaking, an acre of land can host hundreds of thousands[68] of mosquitoes. In ideal habitat (salt marshes and other recently flooded areas), there can be *millions*[69] per acre.

There have been attempts to come up with more definitive answers, however. Mosquito control agencies regularly trap mosquitoes. In a scientific study[70] of the southern Coachella Valley from 1994–95, more than 900,000 female mosquitoes were trapped. Of course, the study didn't trap all the mosquitoes—nowhere close—so there were a great deal more where they came from.

The human population of the *entire* Coachella Valley? Somewhere around 420,000 people[71].

The Worst Party Ever:
An (Involuntary) Blood Drive

So how much blood do mosquitoes take from people on a given night? The answer, of course, varies immensely, but thanks to some baseline research, we can at least hazard a guess for a worst-case scenario.

For example, we know that a mosquito's blood meal usually consists of a few microliters (a microliter is a millionth of a liter). So let's say that a city of 100,000 was holding a (crowded!) block party over ten city blocks (16 acres), and that half the population showed up. Now let's ruin the fun by placing the block party next to a salt marsh, a perfect habitat for mosquitoes. Let's say that conditions are just right, and there are 8 million mosquitoes per acre. That means there are 64 million females.

Even if only 10 percent bit humans, the party wouldn't last long, as there'd be 128 bites per person; if spread out across six hours, that'd make it 21 bites per hour.

The end result? You wouldn't want[72] to go.

If every female fed once and took four microliters, they'd take almost 7 gallons of total blood. For reference, the average human body contains 1.3 gallons. To put it another way, they could fully drain the blood from several people.

Or, if they were virtuous, friendly mosquitoes, they could host a successful blood drive. According to the Red Cross, during a blood drive, each person donates about a pint[73]. The mosquitoes would have collected 54 pints—a pretty impressive total.

Exsanguination: the Worst-Case Scenario

Wild animals aren't the only creatures that are bothered by mosquitoes. Mosquitoes often harm domestic livestock, often leading to significant economic losses. For example, dairy cattle in mosquito-infested areas produce less milk than those in mosquito-free locales. On rare occasions, the problem can get a lot worse—even leading to livestock deaths. Death by blood loss is called exsanguination. That word is usually only found in vampire novels, but in exceptional cases, mosquitoes swarms can actually lead to death by blood loss.

Such worst-case scenarios require special circumstances. In 1980, Hurricane Allen[74] set the stage for just such an event. The hurricane led to widespread flooding in marshes, a perfect environment for mosquitoes. After a week, mosquitoes emerged with a vengeance, killing over a dozen cattle. How did the livestock die? Blood loss after repeated exposure to huge swarms of mosquitoes.

Researchers estimated that it took approximately[75] 3.8 million mosquito bites—5,300 per minute over a 12-hour period—to kill a full-grown steer.

Thankfully, such events are incredibly rare, and humans are essentially immune to this threat because of our greater mobility, not to mention the fact that we live indoors. Other animals (especially smaller mammals) are probably not as lucky.

Pity the Caribou

In Alaska and northern Canada, caribou populations are tormented in the summer months by mosquitoes and other biting insects. At their peak, mosquitoes can be a serious hazard for caribou and cause the caribou to lose significant quantities of blood—reportedly as much as a liter a week[76]. While this rarely kills the caribou outright, it makes surviving on the tundra that much harder.

Watch Out for Grandma Skeeter!

To produce a batch of eggs, most mosquitoes need a blood meal, and most species are capable of producing additional batches beyond the first one. Each batch of eggs has dozens to hundreds of eggs.

Of course, for each additional batch, they require another blood meal from a host. This means that if conditions are right, there can be several consecutive generations of mosquitoes in the air at any given time. So on a given night you could conceivably be bitten by a mosquito *and* its grandmother.

So what about the population of a "killer swarm"? In the summer of 1988, three cattle died in Florida after prolonged harassment by a huge mosquito swarm. Conveniently, we have some idea how many mosquitoes were present thanks to a mosquito trap that was operating just a few miles from the site. It caught thousands of mosquitoes *each night*, including 6,900[77] on just one evening. Plus, that trap was located some distance from the actual kill site, and no trap is anywhere near 100 percent efficient. So as amazing as it might seem, the total number of mosquitoes was certainly much, much worse.

Once Bitten . . .

When mosquitoes produce multiple batches it's not only a nuisance factor, it's also what enables them to carry the most deadly mosquito-borne disease of all: malaria. This occurs because the parasite that causes malaria is not passed[78] to a female mosquito's eggs. Once they emerge as adults, the mosquito must acquire it through a blood meal from an already-infected human. That means that a mosquito requires at least two[79] blood meals to spread malaria: one to contract it and one to spread it.

Other diseases—including West Nile virus[80]—can possibly be passed directly from the female to the eggs, so mosquitoes may be able to spread the disease on the first bite.

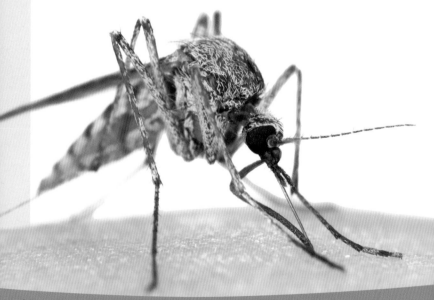

Why Some Years Are Worse Than Others

One of the most common questions about mosquitoes is simple: why are some years worse than others? While there are many variables, the short answer is simple: it usually has to do with weather. The mosquito reproductive cycle depends on water and warm, humid weather. If you have a warm, wet spring, you can expect a lot of mosquitoes right away.

Nonetheless, not all species are found in spring; some are more common in summer. In rare (and terrible!) years, these "seasons" can overlap, giving you two "crops" of mosquitoes at once. And once the mosquitoes are out, they can lay multiple batches of eggs, perpetuating the misery.

In a real respect, a look at mosquito populations is a lot like a retroactive weather report.

The Genetics of Vulnerability to Mosquitoes

If you're really susceptible to mosquito bites, blame your parents. An Australian study surveyed identical twins (who share the same genes) and fraternal twins (who have different genes). The twins were asked about their susceptibility to mosquito bites. The paper's analysis indicated[81] that genetics may be responsible for up to 85 percent of the difference between the two groups. While that data needs to be confirmed, confirmation could go a long way toward explaining why mosquitoes prefer some people over others.

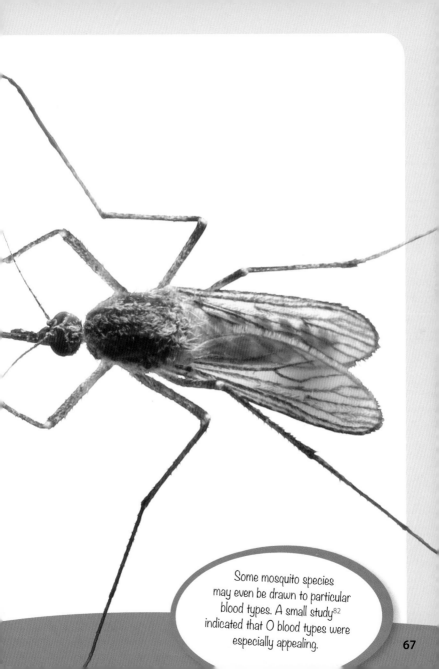

Some mosquito species may even be drawn to particular blood types. A small study[82] indicated that O blood types were especially appealing.

Mosquitoes as Disease Vectors: Flying Syringes

A mosquito bite is bad enough, but mosquitoes have a lot worse to offer: for as long as human civilization has existed, we've been plagued by mosquito-borne diseases, such as malaria, yellow fever and Dengue fever. Despite medical and scientific advances, these diseases still ravage much of the world, killing millions each year.

While North America escapes most of the wrath of yellow fever and malaria, a number of serious mosquito-borne diseases are present on the continent, sickening thousands each year.

To prevent these diseases, one must understand them and the mosquitoes that spread them, so consider this an introduction to the "Flying Syringes" and the diseases they sometimes bear.

Malaria: A Deadly Package Delivered by a Mosquito

Of all the mosquito-borne diseases, malaria is the most widespread and the most deadly. In fact, malaria occupies a dubious place in history. It's almost certainly the most deadly disease we've ever faced, and mosquitoes play a key role in malaria transmission. After all, without mosquitoes, humans wouldn't contract it.

However, malaria is different than most familiar diseases: it's not caused by a bacterium or a virus; instead, it's caused by a parasite (an organism that invades a host and benefits at its expense). The malaria parasites belong to the genus *Plasmodium*, and there are four[83] species that cause the vast majority of cases in humans:

Plasmodium vivax

Plasmodium malariae

Plasmodium ovale

Plasmodium falciparum

Malaria infections all have similar basic symptoms. They feature a roller-coaster combination of fevers and chills, along with sweating, headaches and nausea.

Historically, malaria cases were classified by the severity of their fevers, which were characterized as "benign" or "malignant." Generally speaking, we use a similar standard today, as malaria cases are either described as "uncomplicated" or "severe.[84]"

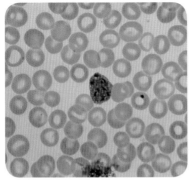

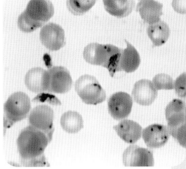

Plasmodium vivax   *Plasmodium falciparum*

While all *Plasmodium* infections are capable of producing serious disease, one species, *Plasmodium falciparum*, is responsible for the vast majority of severe malaria cases and results in the majority of deaths.

Severe cases involve organ failure and neurological problems. Symptoms of "cerebral malaria" include abnormal behavior, seizures and coma[85]. Given that the malaria parasite hijacks red blood cells, it's no surprise that cardiovascular symptoms, including anemia and outright cardiovascular collapse, are common in severe cases, too[86].

Many malaria-related deaths don't occur directly because of malaria; instead, they occur because of secondary infections, such as respiratory infections and kidney disorders[87]. This is similar to AIDS, which also doesn't kill directly; instead, AIDS destroys the immune system and secondary ailments (including the lowly cold) kill the patient.

Attack of the Clones

The life cycle of the malaria parasite has a number of phases, but it's easy enough to summarize. Mosquitoes carrying the parasite bite a human, injecting some of their saliva with the bite. This transfers sporozoites, which are basically "seed parasites," to the host. These migrate to liver cells where they reproduce asexually, producing trophozoites (immature cells) and then schizonts, which are blob-like clusters essentially full of parasite clones (called merozoites). The schizonts rupture, releasing the clones, which then begin invading red blood cells.

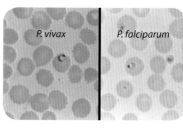

P. vivax   P. falciparum

Once the parasites have entered the red blood cells, two different things happen: some clones continue reproducing, and rupturing, ensuring a steady supply of additional merozoites. In the process, they destroy the cell, consuming[89] its hemoglobin. This produces granules with a characteristic black-brown pigment; this pigment is visible under a microscope and is the hallmark of the disease.

Other red cells produce gametocytes, the precursor to the parasite's reproductive cells. These don't cause symptoms but are ingested by mosquitoes, and thereby perpetuate the parasite's life cycle—and the spread of malaria.

When Europeans colonized Africa, most lacked any immunity for malaria and many died. This led Africa to be deemed "White Man's Grave." Malaria targets the poor and the rich alike; it has killed countless peasants but likely contributed to the death[90] of King Tut and may have even killed Alexander the Great[91].

● Malaria risk
● No malaria

A Terrible Toll

Around 150–300 million people died from malaria[92] during the twentieth century. Believe it or not, historically speaking that was a *good* hundred years, as the century saw worldwide mortality rates plummet. For a frame of reference, at the turn of the century, malaria was quite common. In the U.S., hardly a malaria hotspot, over 8,900 people died[93] from 1900–04. Today, deaths in the U.S. are unheard of, even though over a thousand people a year contract it, usually after traveling to malaria-infested areas.

Other parts of the world—especially sub-Saharan Africa—aren't so lucky. According to the World[94] Health Organization, there are somewhere around 660,000 deaths due to malaria a year—and new research[95] indicates that this number may be wildly understated and that there may be as many as 2 million deaths per year.

Humanity and Malaria: Familiar Foes

Malaria has plagued much of humanity since the beginning of civilization. In fact, the creation of widespread agriculture—and civilization itself—almost certainly[97] made us one of the primary sources of the parasite. The story is pretty simple: prior to the advent of civilization, humans were primarily hunters and gatherers, but once people began practicing agriculture, everything changed. We began living in villages, towns and cities, and we began clearing land. This created habitat for mosquitoes, and once they were born, they no longer found the variety of prey they were accustomed to—instead, they found humans and adapted to feeding on us. Some malaria-carrying species now feed primarily, or in some cases, almost entirely[98], on humans. In a real respect, malaria followed agriculture.

This also explains why malaria is less common in many Western countries. As populations have become more centralized in urban areas, they have essentially moved away from prime mosquito habitat; there simply aren't as many places for mosquitoes to develop in a city. Widespread mosquito control programs don't help their chances, either.

Malaria predates humanity, and a *Plasmodium* parasite (which infected birds) has even been found in Dominican amber[99].

Evolution in Action:
the Malaria Hypothesis

Even though much of humanity has been under assault by malaria for millennia, some human populations have evolved genetic defense mechanisms, adaptations that are specifically tailored to combat malaria, or at least make it less deadly.

A number of examples exist, but the most famous causes sickle cell anemia, a disease that is present in a significant portion of some African populations. It causes red blood cells to appear crescent shaped, akin to a sickle or a crescent moon. In regions with widespread malaria, this adaptation has its advantages, as it reduces the likelihood of death by malaria by up to 90 percent[100].

Unfortunately, such adaptations have serious reproductive costs. Sickle cell anemia is a potentially deadly disease, and without treatment, serious cases are often fatal; even when it's not fatal, it can often lead to significant[101] medical problems, including stroke, and it shortens life expectancy considerably.

But in the most malaria-prone areas, many children succumb to malaria before they are four[102]. That's the trade-off of natural selection: in terms of big picture, more people with the sickle cell trait avoid the worst ravages of malaria when they are children and live to pass on the gene.

Some people also simply carry the sickle cell trait without having the full-blown disease. For a child to get it, both parents need to have it. For this reason, the disease is found in a small subset of the African-American population in the U.S.

Quinine and Sweet Wormwood Save the Day

For most of history, humanity had little defense against malaria. It wasn't until the seventeenth century that an effective anti-malarial—quinine (pronounced *KWY-nine*)—first found widespread use. Long used by the indigenous Quechua tribes of Peru, quinine soon became the drug of choice against malaria.

Eventually, however, malaria began to show tolerance to quinine (and even the other drugs that were later based on it). Thankfully, another treatment appeared in the 1970s. Known in China as an herbal medicine for over 1,600 years, the sweet wormwood plant (*Artemisia annua*) contains a compound known as artemisinin[103], which was isolated in the 1970s and is now the basis for many highly effective anti-malarial medicines.

Travelers visiting areas with widespread malaria now routinely take quinine- and artemisinin-based drugs to prevent infection.

The Cinchona plant, the source of quinine

Quinine is also an ingredient in tonic water, the bitter component of a gin and tonic. Tonic water contains much less quinine that a medicinal dose, however. (Oddly enough, because of the presence of quinine, tonic water is also fluorescent under a black light.)

Efforts to Eradicate Malaria

On the whole, the last century saw remarkable progress in combating malaria, and health organizations even had the audacious goal of eliminating the disease globally. They had some success, and malaria was virtually eradicated from large parts of the world. Unfortunately, it wasn't eradicated everywhere, and in some areas, especially areas of Africa, things have actually gotten a lot worse. From 1980–2004, malaria deaths in children rose threefold. In 2004, there were a horrifying[104] 1,047,000 deaths in children under five years of age. Worse yet, parasites have started to demonstrate resistance to the existing drugs.

Nonetheless, recent efforts have helped reduce malaria deaths significantly. Bed nets treated with insecticides are now being distributed widely and help prevent nighttime exposure to mosquitoes. New preventative medicines and treatments are also being developed, including the Holy Grail of malaria research: an effective vaccine. A number of projects are in the works, including a vaccine that is now being tested in humans by the U.S. National Institutes of Health. That vaccine consists of malaria parasites that have been weakened, enabling the body to learn how to eliminate them. The vaccine has shown[105] some promise—when the highest doses were administered the greatest number of times, it had a perfect protection rate. That doesn't mean we've solved the problem, however; unlike most vaccines, it's delivered intravenously (directly into a vein) and it still needs to undergo a great deal of additional testing in order to determine how long its protection lasts.

Yellow Fever and Dengue Fever

Mosquitoes don't just spread malaria; they also spread yellow fever and Dengue fever. Unlike malaria, yellow fever is a virus, and it isn't transmitted by *Anopheles* mosquitoes. Instead, *Aedes* mosquitoes are some of the primary carriers. Found throughout a large portion of Africa and South America, yellow fever gets its name because it leads to liver damage and jaundice, which causes skin (and eyes) to look yellow. While most cases are moderate, severe cases involve jaundice, organ failure and bleeding, and have a high fatality rate—somewhere between 20–50 percent[107]. Thankfully, a vaccine exists, and travelers are recommended to get it before visiting areas where the disease is present.

Dengue (*den-GHEE*) fever is a widespread disease caused by a family of four related viruses and is spread via *Aedes* mosquitoes. Dengue fever is found throughout much of the tropics and the subtropics, including the southernmost portions of the U.S. It can't be transmitted from person-to-person, but travelers can become infected overseas and return home with the virus, where native mosquitoes can then bite them, thereby spreading the virus. Infection leads to a serious fever and headache, and can progress to Dengue[108] hemorrhagic fever, which can lead to rapid blood loss, shock and death. No vaccine exists, but one[109] is in development. Thankfully, prompt care significantly improves the likelihood of survival.

West Nile Virus: New Kid on the Block

A relative newcomer as far as mosquito-borne diseases in North America go, the first U.S. case of West Nile virus was reported in 1999. Since then, there have been outbreaks of the disease each summer, with over 16,000 confirmed cases in all—and more than 1,500 deaths[110]. Mosquitoes feed on infected birds and then bite humans, transmitting the disease, which is characterized by a fever, headache, fatigue and various aches and pains. Serious cases affect the brain, causing high fever, meningitis/encephalitis, and in the worst cases, death.

The first cases of West Nile were discovered in New York City[111], and within a few short years, it spread across much of North America. It was able to conquer most of the continent thanks to its unique transmission pattern. Birds are the primary reservoir of the disease, and they are infected by mosquitoes, which transmit the disease. Those birds then migrate, bringing the virus with them. This process repeats, helping the virus expand its range and survive year-round.

Mosquitoes and Encephalitis

Mosquitoes also transmit a number of viruses that cause encephalitis —a disease that leads to swelling of the brain. Examples include St. Louis encephalitis and the dastardly duo of Western equine encephalitis and Eastern equine encephalitis, which can affect humans as well as horses and a variety of other animals. These diseases are generally more common in the summer (when mosquitoes are most active). All share somewhat similar symptoms—most cases are mild and lead to only a mild fever and a headache. Severe cases can lead to brain swelling, seizures and death; a significant number of survivors often have long-term brain damage[112]. As there are no vaccines for these diseases, bite prevention is especially important.

Where does St. Louis encephalitis get its name? While the disease now occurs around the country, the first recorded outbreak of the disease occurred near St. Louis, Missouri.

Heartworm

Humans aren't the only ones who have to worry about mosquito-borne illnesses. Mosquitoes infect a variety of other animals, including apes, monkeys, dogs, cats and even sea lions[113]. The most familiar scourge is heartworm, which is a parasitic worm that often infects dogs. When dogs are bitten by mosquitoes bearing the parasite, heartworms take up residence in the dog's body—especially the heart and lungs—growing and reproducing all the while. Male heartworms can even reach a somewhat astounding (and disgusting!) foot[114] in length. In the worst cases, the heartworms literally are so thick that they block blood flow, leading to heart failure and death.

Thankfully, preventing heartworm infection is easy thanks to low-cost medications like ivermectin, which are administered in monthly doses and kill heartworm larvae. Treating for the entirety of heartworm season (which varies by area) makes it next to impossible for heartworms to develop, even if heartworm-carrying mosquitoes bite a dog.

Avoiding Mosquitoes Made Easy

The best way to avoid mosquito bites is to stay away from them altogether. Unfortunately, unless you're a recluse, that's not always possible. Thankfully, there are many ways to minimize your exposure to them. To do so, you need to know when mosquitoes are most active, wear the proper clothing outdoors and select an effective repellent to keep the skeeters away.

Given how common mosquitoes are—and how many products exist to counteract them—doing something as simple as choosing a mosquito repellent can be daunting. The following tips will help you avoid peak-mosquito times, make you less attractive to the mosquitoes you encounter, and help you discover expert-recommended and effective repellents. Along the way, I'll also de-bunk some of the superstitions and bogus mosquito cures that are constantly repeated.

Get Rid of That Standing Water!

If you hate mosquitoes, make sure your yard isn't inadvertently giving them a place to reproduce! While all mosquitoes require water to reproduce, they don't need much, and they reproduce just as easily in the old tires in your backyard as in a seaside marsh. So the first step to fighting mosquitoes is to look around for places that can collect water—everything from the stagnant water in the birdbath to clogged-up gutters and the tarp covering the boat in the backyard. Dumping the water out will kill any larvae present, and removing the container will force female mosquitoes to look elsewhere to lay their eggs.

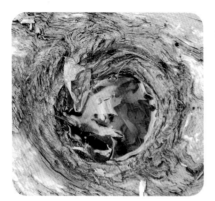 

While you're dumping out the water, fix your window screens. Even a small puncture is enough for mosquitoes to make their way into your home—and several disease-carrying species have no qualms about doing so. Repair kits are cheap and easy to apply, and can make a world of difference if you are in an area with a lot of mosquito activity.

If you have a birdbath, consider buying a "water wiggler." These devices are cheap and agitate the water enough to prevent mosquitoes from laying their eggs there.

Location, Location, Location
—and Time

Like the DMV, political ads, and other less pleasant facts of life, mosquitoes are here to stay. Nonetheless, if you know something about their habitat and behavior, you can avoid the worst the swarms have to offer.

Mosquitoes can be found almost anywhere, but some areas are generally much worse than others. Ideal mosquito habitats include heavily wooded areas or areas with lots of vegetation, especially if they are near stagnant water. Examples include floodwater areas, tidal marshes and swamps.

Mosquito species are also often active at specific times of day. While some species—including the Asian tiger mosquito—bite all day long, most mosquitoes rest in cool, shaded areas during the heat of the day and venture out in the afternoon, evening, and early morning.

These two tips are basically cumulative—if you avoid ideal mosquito habitat and are active when they usually aren't, you're far less likely to encounter them. On the other hand, if you decide to take a summertime after-dinner hike near a river, you might be running back to your car.

What's a Mosquito's Favorite Color?

There really is an answer—black. In a number of studies[115], mosquitoes were drawn more to hosts that were wearing darker-colored clothing. Black was the most popular, but reds and blues often made the list. Subjects wearing lighter-colored shirts —especially white—were targeted less often.

The moral of the story? Easy! Wear light-colored clothing when in mosquito territory, and when you're picking out your outfit, find something long-sleeved but loose-fitting. While mosquitoes can bite through some clothing, they have a lot more trouble if that clothing is loose, as it adds distance between your skin and the mosquito. A loose outfit also helps the body trap less heat, making it more difficult for mosquitoes to find you.

DEET: The Gold Standard

DEET is the most commonly used insect repellent. Invented in the 1940s by the U.S. Army, the name "DEET" is an abbreviation for (the incredibly long!) chemical name N,N-Diethyl-3-methylbenzamide.

By itself, DEET is a yellowish oil, but we're most often familiar with it in a variety of different mediums, including lotions, creams and sprays. Depending on the product, DEET concentrations can vary significantly. Some containers of insect repellent have a low concentration of DEET, whereas others contain much more. (Always consult with a doctor or pharmacist when choosing a concentration.) More DEET gives you longer protection, but at a certain point—above 30–35 percent—this benefit plateaus.

While it's not entirely clear why DEET works, current research indicates[116] that repellents interfere with the mosquito's sense of smell: either by blocking the specific appealing odors we emit or scrambling the mosquito's fine-tuned sense of smell, which it depends on to find a blood meal. What's more, according to one study, DEET doesn't simply repel mosquitoes in flight—it also repels those that land, but for an entirely different reason: it seemingly tastes bad[117].

General Tips For Using DEET[118]

Follow the directions on the label.

- Don't use more than you need—use just enough to cover your skin and clothing.
- DEET melts plastic, so be careful when applying it.
- Apply DEET outdoors.
- Do not put on open cuts or injuries, and don't use beneath clothing.
- Consult with a doctor to determine which concentrations to use on children, and don't use on very young children without a doctor's OK. Apply DEET for children.
- Also, don't spray DEET (or any substance, really) into your face or your nose, eyes or mouth. Instead, spray some into your hands to apply it.
- Once you're indoors, wash any remaining DEET off, and wash DEET-exposed clothing.
- Store insect repellents safely and out of reach of children.

There are downsides to DEET. Its texture and smell are decidedly unpleasant, and it also has the rather disconcerting tendency to melt plastic and rubber (really!).

While some have qualms about its safety, it is widely recommended[119] by doctors and health[120] authorities alike. The most common complaints are mild[121] skin reactions (rash, redness) and problems when users inadvertently spray DEET into their eyes.

Pyrethrum, Pyrethrins and Pyrethroids

Chrysanthemums to the Rescue

DEET may have been invented in a laboratory, but many insect repellents are inspired by the natural world. Why? Humans aren't the only species that practice chemical warfare; in fact, chemical defenses are incredibly common in nature. From the compound in catnip that drives cats[123] crazy to the citronella oil in the countless torches that burn in backyards in summer, many insect repellents or insecticides are derived from, or inspired by, insect repellents that already exist in nature.

Believe it or not, one of the finest naturally inspired mosquito-killing compounds comes from a plant you're probably already familiar with—the chrysanthemum.

Long a favorite of gardeners, the flowers of certain chrysanthemum species produce an insecticide that is referred to as pyrethrum (*pie-REETH-rum*). It contains compounds that are called pyrethrins (*pie-REETH-rins*), which are highly effective insecticides.

Until the middle of the century, producing flower-based insecticides was problematic, as manufacturers were at the mercy of the growing season. (After all, they needed plants to produce the repellent!)

Plus, when the repellent was actually used, the products degraded in light fairly quickly. Thankfully, synthetic equivalents called pyrethroids (*pie-REETH-roids*) were produced[124]. These naturally inspired[125] products can be produced in a factory and are therefore available year-round. Thanks to slight alterations, they also are safer[126] for mammals and last longer when in use.

Today, there are a number of pyrethrum-inspired insecticides, but one of the most common is called permethrin (*per-METH-rin*).

A dried chrysanthemum flower

Permethrin is for Use on Clothing, Not Skin

Unlike DEET, permethrin is not a topical repellent and shouldn't be used on skin. Why? Two reasons. First, the body quickly metabolizes permethrin, so it'll likely wear out by the time you need it. More importantly, use on skin can also cause a mild reaction[127]. Permethrin is also more of an outright insecticide than it is a repellent, though it does have some repellent effect. When a mosquito encounters the permethrin on clothing, it is often killed outright.

Thanks to its efficacy against mosquitoes and other problematic critters (including ticks!), permethrin-infused clothing is popular. This is why the U.S. military now issues permethrin-dosed uniforms in areas with mosquito-borne diseases. If properly washed, the clothing retains the permethrin without exposing the soldier to too much of the compound[128]. Permethrin-impregnated clothing is also available online, and spray-on products are available as well, making it easy to make any clothing insect-resistant. Generally speaking, sprayed-on permethrin will remain effective for six washings.

All in all, permethrin is pretty useful stuff. When permethrin is used on clothing and a topical repellent[129] like DEET is used on skin, they can provide almost complete[130] protection from mosquitoes.

General Tips For Using Permethrin on Clothing[131]

- Remember, permethrin should be used on clothing, not skin.

- Always follow the directions on the label and choose the correct type of permethrin product. When in doubt, consult a doctor.

- Apply the permethrin outside, and don't let children apply the spray to clothing. Consult with your doctor before using it on children's clothes.

- It may seem obvious, but don't apply permethrin to clothes while you are wearing them.

- Once applied, permethrin products need to dry thoroughly. Hang the clothing outside to dry.

- You can wash permethrin clothing, but make sure to wash permethrin clothes separately.

- Store insect repellents and insecticides safely and out of reach of children.

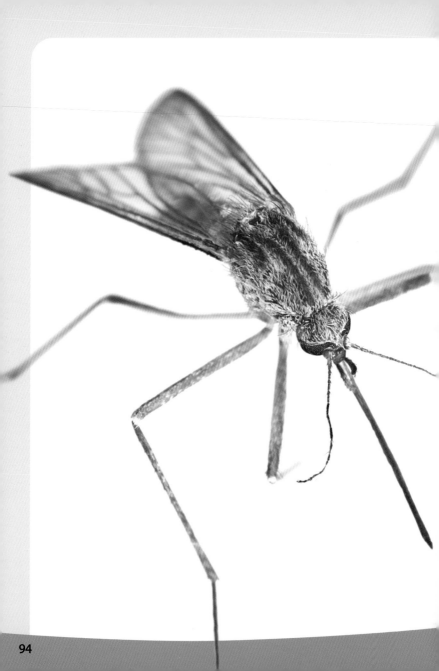

Using Other Types of Permethrin Insecticides

As a broad-spectrum insecticide, permethrin isn't just used for clothing. It is also found in products used in everything from municipal mosquito control to sprays for pets and livestock.

- As there are many different permethrin products, always read the product labels carefully, and make sure you select an appropriate product for your use.

- Permethrin is a broad-spectrum insecticide, which means that it kills a wide variety of insects. If you "fog" your yard, be aware that you're not just killing mosquitoes; you're also killing species you may want to keep around, such as bees and fireflies.

- Permethrins cause serious damage to aquatic environments, so be very careful when using them around streams, ponds, or other wetlands.

- If you're wary of using permethrin on your yard, contact your local extension service for other insect-control options.

The next time the mosquitoes chase you out of your backyard, take comfort in this: when initially applied, permethrin is known for its "knockdown effect." It's exactly what it sounds like; when insects come into contact with it, they are knocked to the ground and lose control of their nervous system, often leading to spasms. According to a pair of researchers at the University of Montana, the effects can be dramatic—causing the mosquito to even lose its wings and legs[132]!

Picaridin

Another Product Inspired by a Natural Repellent

In the U.S., picaridin[133] is a relatively new weapon in the fight against mosquitoes and other troublesome insects. Available since 2005, it is a synthetic repellent inspired by piperine, the chemical that gives black pepper its characteristic smell. Historically, pepper has been used as an insecticide. In field tests, picaridin has shown to be as effective or nearly as effective as DEET, without many of the negative side effects. Unlike DEET, picaridin doesn't melt plastic, and it doesn't feel greasy or have DEET's (terrible!) smell. Studies have indicated that it is also less toxic than DEET, though there is not much data about long-term exposure to picaridin.

General Tips For Using Picaridin[134]

- Follow the directions on the label.

- Don't use more than you need—use just enough to cover your skin and clothing.

- Apply picaridin outdoors, and don't let children apply the product themselves. Just to be safe, consult with your doctor to determine which concentrations to use on children.

- The Mayo Clinic recommends[135] against using picaridin on children under three.

- Do not put on open cuts or injuries, and don't use beneath clothing.

- Also, don't spray picaridin into your face or your nose, eyes or mouth. If you want to apply it to your face, spray some into your hands to apply it.

- Once you're indoors, wash any remaining picaridin off, and wash exposed clothing.

- Store insect repellents safely and out of each of children.

Eucalyptus Isn't Just for Koalas

Oil of Lemon Eucalyptus, a Natural Mosquito Repellent

If you're looking for a natural mosquito repellent that is effective, take a page out of the koala's playbook. The lemon eucalyptus tree (*Corymbia citriodora*), which is native to Australia, happens to be one of the koala's favorite snacks. It is also one of the sources of oil of lemon eucalyptus, which is a potent insect repellent. Studies indicate that it can be just about as effective as DEET and lasts about as long. "PMD" is the abbreviation for the chemically synthesized version of the same substance, and because it is easier to produce, it's often found in many repellents. (The actual oil of lemon eucalyptus must be obtained from trees.)

Oil of lemon eucalyptus has a few other benefits: it's nontoxic, it has a pleasant scent, and it lacks some of DEET's side effects (melting plastic!) [136, 137, 138].

General Tips For Using PMD and Oil of Lemon Eucalyptus

- Follow the directions on the label. **Note:** do not use this product on children under three.

- Do not use "pure" oil of lemon eucalyptus, which is sometimes labeled as "essential oil of lemon eucalyptus." The CDC notes that it has not been tested for safety[139] or efficacy.

- Don't use more than you need—use just enough to cover your skin and clothing.

- Apply outdoors, and don't let children apply the product themselves.

- As with any chemical, don't get this product in your eyes; eye irritation can result.

- Do not put on open cuts or injuries, and don't use beneath clothing.

- Also, don't spray it into your face or your nose, eyes or mouth. If you want to apply it to your face, spray some into your hands to apply it.

- Once you're indoors, wash any remaining repellent off, and wash exposed clothing.

- Store insect repellents safely and out of reach of children.

IR3535

A Clunky Name, but a Popular Choice Across the Pond

"IR3535" is the trademark name for a chemical compound that is similar to the naturally occurring amino acid called B-alanine[140]. Widely used in Europe for the past 20 years, IR3535 was introduced to the North American market only relatively recently. Testing indicates that it's reliable and effective. More importantly, it's also been deemed nontoxic[141] by the EPA; in its long history of use overseas, there have been no safety concerns raised about its use.

General Tips for Using IR3535

- Follow the directions on the label[142]. Be wary of products that combine sunscreen and IR3535, as sunscreen needs to be applied more often than most bug sprays, leading one to apply too much repellent.

- Like DEET, IR3535 can melt plastic, so take caution when applying it.

- Don't use more than you need—use just enough to cover your skin and clothing.

- Apply outdoors, and don't let children apply the product themselves.

- As with any chemical, don't get this product in your eyes; eye irritation can result.

- Do not put on open cuts or injuries, and don't use beneath clothing.

- Also, don't spray it into your face or your nose, eyes or mouth. If you want to apply it to your face, spray some into your hands to apply it.

- Once you're indoors, wash any remaining repellent off, and wash exposed clothing.

- Store insect repellents safely and out of reach of children.

What Doesn't Work

Superstitions, Mosquito Traps and Other Bogus "Cures"

Whether it's Avon's Skin So Soft or DEET-covered wristbands, it seems like everyone knows of a favorite trick to avoid mosquitoes. Unfortunately, beyond the tried-and-true repellents, many of the "you-won't-believe-this-works!" claims are just that—unbelievable. Here are a few of the worst offenders mentioned online or in chain emails. This is by no means a complete list—new ones keep popping up every year—but it's a window into the world of bad science.

Avon Skin So Soft

Avon Skin So Soft bath oil is often claimed to be a mosquito repellent, and according to a study published by the *New England Journal of Medicine*, it *does* repel some mosquitoes. The problem? It only works for a few minutes—according to the study, it only provided complete protection for an average of nine[143] minutes, compared to over 300 for DEET. Adding to the confusion, Avon Skin So Soft now produces lotions that contain a legitimate insect repellent (IR3535), so those are far more likely to work than the plain-old bath oil.

DEET-covered Wristbands

The idea behind DEET wristbands seems simple enough: use a well-known and well-proven repellent and add it to wristbands. In theory, a "cloud" of DEET would surround the user, protecting against mosquitoes. The problem? They don't work very well;

instead of protecting one's entire body, the wristbands only protect in the immediate vicinity of the wristband. Mosquitoes have no problem landing—even nearby—as there is no DEET present to repel them.

Vicks VapoRub (and Listerine)

Most people associate Vicks VapoRub with head colds and sick days, but some people—especially on the Internet—claim that it's an excellent mosquito repellent. This isn't true, however. Vicks VapoRub consists of three active ingredients: camphor, menthol and eucalyptol, along with a number of inactive ingredients. All three of these compounds are found in low concentrations (single digits, percentage-wise).

Part of the confusion may come from the fact that the oil from certain eucalyptus species is a potent insecticide and mosquito repellent, but despite its similar-sounding name, eucalyptol is a different compound. While there have been some tentative studies showing that eucalyptol[144] can be an effective repellent, those studies used concentrations that were dozens of times higher and consisted of up to 75 percent eucalyptol. Vicks VapoRub contains just 1.2 percent of the stuff. Similar claims are made about Listerine, but it, too, contains only a tiny amount of the compound, certainly not enough for any sustained amount of protection, if it exists at all.

Marigolds

While many mosquito repellents are inspired by plants—and even common flowers like the chrysanthemum—there's no evidence that planting marigolds will affect mosquito populations. This myth is intertwined with another claim about marigolds—that they protect gardens against harmful insects. While in very specific situations they can protect gardens against nematodes, marigolds can actually attract harmful mites. As anyone who has been hounded by mosquitoes amid their marigolds knows, sadly they don't have a repellent effect on mosquitoes, and no repellent effect has been documented in any study.

Bananas, Onions, Garlic[145], Vitamins

Various foods are often claimed to be mosquito-attractants—or alternatively, repellents. As much as it would be fantastic to avoid mosquitoes by indulging in a giant plate of garlic fries, the truth is a lot more complicated. Mosquitoes are attracted to a very complex set of environmental cues, including heat, carbon dioxide and hundreds of compounds in the breath and skin. While some products, such as beer, *can* actually attract mosquitoes in some cases, hard-and-fast examples like this are very, very rare in the scientific literature. So unless you're looking in the insect repellent section, it's pretty unlikely you'll find the ultimate mosquito deterrent at the grocery store.

Bat Boxes and Purple Martins

 Bats are lean, mean insect-eating machines. Impressive predators, they use sonar to hone in on their prey, so it only seems logical that adding to the bat population would decrease the mosquito population. The problem? Bats don't actually eat all that many mosquitoes, instead preferring larger prey, such as moths[146], which provide more nutritional value. While they may decrease the mosquito population in an area by a small amount, bats certainly aren't a reliable control measure.

Purple martins are a similar case. While it's true that they eat insects, they don't primarily eat mosquitoes. On the contrary, they tend to eat[147] a wide variety of different insects, including dragonflies, a mosquito predator.

Mosquito Traps: Don't Buy the Hype Just Yet

Mosquito traps are currently one of the more popular tools to control mosquito populations. Generally speaking, these products lure mosquitoes to the trap—usually by emitting carbon dioxide or some other "lure." When the mosquito arrives, it is usually trapped or killed (either via suction, a fan, or by being zapped). While the technology shows some promise, they are nowhere near a guaranteed mosquito control solution. After all, there are over a hundred mosquito species in the U.S., and not all species are attracted to the same "lures," so some species may get trapped while others are unaffected. What's more, even if the traps catch a good number of mosquitoes, this doesn't necessarily mean it will make a noticeable dent in the mosquito population in your area. (This is especially true if you live near prime mosquito habitat, or if mosquito populations are entering your area from elsewhere.)

Bug Zappers

While mosquito traps show some promise, bug zappers are next to worthless when it comes to mosquito control. These devices usually operate by luring unsuspecting insects toward a light that is attached to an electrified surface. When the bugs land, they complete the circuit and are electrocuted. The problem is, mosquitoes are not drawn to such zappers very often; instead, many non-pest species (including many types of beetles and moths) are inadvertently killed, harming the local ecosystem.

Sonic and Ultrasonic Devices

Simply put, ultrasonic devices claim to repel mosquitoes simply by using sound—either by mimicking the sound of a prey species or by discouraging already-pregnant females from landing by imitating the sound of a male mosquito. Such devices can range from keychain-sized devices to speaker-sized machines, and they operate either in the audible sound range or in the ultrasonic range (which we can't hear). All such products share one feature in common: they don't work. They've been tested and evaluated[148] for over a decade[149], and they are next to worthless.

The Problem with Natural Cures

When given the choice, many people seem to prefer natural insect repellents rather than synthetic ones, and there are now many "all-natural" insect repellents on the market to choose from. The problem is, most simply don't work as well as synthetic repellents, such as DEET.

Many natural compounds—including some found in all-natural bug sprays—do have a repellent effect, but that doesn't mean they make usable repellents. After all, the repellent has to be concentrated enough to repel mosquitoes but low enough to be nontoxic. What's more, just because a natural product is safe and nontoxic doesn't mean that it will work for long. As things stand, very few all-natural products are as effective, safe and dependable as synthetic repellents such as DEET.

When it comes to mosquito repellents, the fear of chemicals can literally be dangerous, as it can lead people to use a less-effective product. This can make mosquito bites—and potential disease transmission—more likely.

Mosquito Control
on the Home Front

Mosquito Dunks and Citronella

For the average person, two types of mosquito control are especially popular—mosquito dunks and citronella torches.

Mosquito dunks are small discs that resemble small hockey pucks. They are essentially bacteria "starter kits" and contain the bacterium *Bacillus thuringiensis israelensis*, which is nontoxic[150] for humans but lethal for mosquito larvae. These bacteria thrive for a short period in water (usually around two days), but in the process they poison any mosquito larvae they encounter. (Talk about a satisfying image: they literally cause the larva's stomach to explode.) Dunks can certainly help eliminate some mosquito larvae in areas with standing water, but they aren't a cure-all by any means.

On a summer's night, bamboo citronella torches are a common sight, and citronella-based lotions and sprays are common as well. While some formulations of citronella oil can work against mosquitoes, they often have to be applied quite often (every two hours) and efficacy varies a lot by formulation. Citronella torches have an even spottier track record. In studies, they do have some repellent effect, but they don't offer anywhere near complete protection[151], so don't depend on either as your only line of defense.

A Mosquito Miscellany

The Weird, the True, and the Funny of Mosquito Lore

Mosquitoes have been with us since the dawn of civilization, so it's not surprising that they are common symbols in popular culture. T-shirts emblazoned with mosquitoes boast that they are New Jersey's State Bird or Wisconsin's Air Force, and an anthropomorphic mosquito—Skeeta—even runs in mascot races between innings at

Minnesota Twins games. (Tellingly, she was born in Everywhere, Minnesota.) From Air Force bombs full of mosquitoes to the World Mosquito Killing Championship in Finland, here's a brief review of some funny and strange aspects of mosquito lore.

Mosquitoes in War

In most conflicts in human history, malaria and other mosquito-borne diseases have played as an important role as the combat forces themselves. This was even true in World War II, when hundreds of thousands of U.S. troops were infected with malaria, and quinine was in short supply. The situation got so bad that General Douglas MacArthur famously[152] remarked: "This will be a long war if for every division I have facing the enemy I must count on a second division in hospital with malaria and a third division convalescing from this debilitating disease!" He thereafter made malaria prevention a priority, which helped reduce the number of cases and made the combat force more effective.

The mosquito threat was so real that American propaganda actually personified malaria and the most common malaria-spreading mosquitoes: Malaria Moe was featured in a cartoon and the *Anopheles* mosquito was known as Anopheles Ann. She was illustrated by none other than Dr. Seuss.

Bugs in Battle:
Entomological Warfare!

Just as mosquitoes have plagued militaries the world over, many attempts have been made to weaponize insects in so-called entomological warfare. From the 1950s to the early 1980s, the U.S. military conducted a significant amount of research into using bugs rather than bombs, and they spent a good deal of time attempting to weaponize the yellow fever mosquito (*Aedes aegypti*). These plans weren't hypothetical; several large-scale operations were carried out to test the idea. In one, Operation Big Buzz (really!), several hundred thousand uninfected mosquitoes were released over an area of Georgia, and this operation was followed up with others that released mosquitoes over populated areas[153]. The tests indicated that yellow fever mosquitoes and other biological agents were likely to be an effective weapon of war; terrifyingly, one part[154] of the report even included a "cost per death" chart. Thankfully, the programs have since been scrapped.

In the movie *Jurassic Park*, mosquitoes that fed on dinosaurs are found encased in amber. The dinosaur material is then used to resurrect the dinosaurs. (Spoiler alert: Things don't go well after that.) While this may seem plausible, it doesn't work in practice, as DNA is fragile, and a good deal of amber formed after dinosaurs had become extinct. Worse yet, in the movie, the mosquito species[155] shown in amber didn't actually ingest blood to begin with.

Time for Payback

The World Mosquito Killing Championship

If you hate mosquitoes, you should book a ticket to Finland so you can visit tiny Pelkosenniemi, where the World Mosquito Killing Championship is held each year. The rules are simple: Whoever can kill the most mosquitoes in five minutes by hand wins. The current record holder is Henri Pellonpää, who killed 21 mosquitoes, eclipsing[156] the previous record of 7.

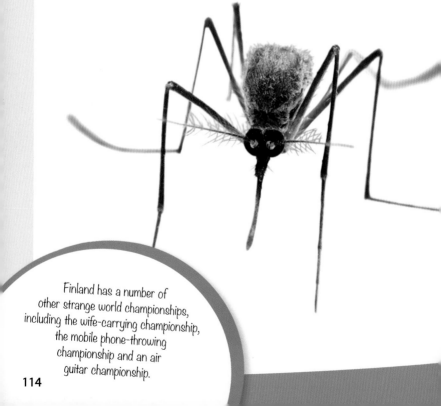

Finland has a number of other strange world championships, including the wife-carrying championship, the mobile phone-throwing championship and an air guitar championship.

Annoyance, Meet Terror: The Botfly

While mosquito-borne diseases pose a significant disease risk, mosquitoes in North America are often simply an annoyance. But for all of the countless outdoor evenings they spoil, it could be worse. We could have mosquitoes *and* botflies. At first glance, botflies almost look cute—a bit like a bumblebee. Once you learn a bit more about them, they don't look so cute.

Mosquitoes (and other biting insects/arthropods) play a central role in the botfly's life cycle; female botflies capture mosquitoes and deposit eggs on them. When the mosquito bites a host, the larvae hatch and make their way into the wound, where they anchor themselves *under the skin* and grow. This produces a noticeable bump; eventually the larva emerges, much larger, where it pupates, becoming an adult[157] several weeks later.

Where do the host species lay their eggs? In all sort of large warm-blooded mammals, including humans. Don't get worried just yet. Cases are almost unheard of in the U.S., and even where botflies are found, human infections are mercifully rare. Even if the presence of a botfly larva is suspected, there are many ways to remove it. One can block the larva's breathing tubes (spiracles) with petroleum jelly or wax or another thick substance, even bacon[158]! You can also remove them by literally squeezing them out (ouch!), but this raises the risk of infection. Thankfully, relatively minor surgery often works just fine.

An Ongoing Science

Over the last two centuries, researchers have made great strides toward understanding mosquitoes and mosquito-borne diseases, but there are many questions that still remain. In a certain respect, this should be expected—there are thousands of mosquito species worldwide and dozens of mosquito-borne diseases. The subject matter is inherently complex, and we're still dealing with some basic questions. Case in point: A large swathe of disease-bearing mosquitoes—which had previously been considered part of the *Aedes* genus—was recently reclassified into an entirely different genus[159]. While that's a technical change, it's a pretty important one, and it goes to show just how much we have to learn.

New Approaches and Treatments

At any given time, a number of treatments for mosquito-related diseases are under development and new mosquito repellents and barriers are being created.

Some of these treatments seem to have quite a bit of promise. Vaccines for malaria and Dengue fever are undergoing early testing, and at the same time, other repellents are being developed.

One seemingly promising repellent is the Kite Patch, which is based on research from a major study in the journal *Nature*[160]. As its name suggests, the Kite Patch is just that—a patch that emits chemicals that are alleged to make wearers invisible to mosquitoes. This makes the Kite Patch a spatial repellent; if it works, the days of slathering on DEET may be over. While initial reports seem legitimate, the patch still needs boots-on-the-ground testing. Thankfully, that's underway, and a large-scale test will be launched shortly in Uganda, one of the most malaria-infested countries in the world. If successful, it could put a serious dent in the number of cases of tropical diseases.

Genetic modification is also likely to serve as a new weapon against mosquitoes. A number of proposed modifications have been suggested—everything from releasing sterile mosquitoes to modifying mosquitoes to make them immune to malaria.

Invasive Species and Mosquitoes in a Warming World

Just as there are many remaining scientific questions about mosquitoes, new questions are arising thanks to climate change. Some of the questions are basic—how much will mosquito habitat be affected by climate change, and what will happen? Will mosquito species inhabit areas where that species wasn't found before? If so, will they bring diseases like yellow fever or malaria with them?

The answers are unlikely to be simple—mosquitoes are complex! —but given the reality of climate change, these are certainly questions to study.

Invasive species are a concern, too. The Asian tiger mosquito was accidentally introduced to the U.S. in 1985, and it is now common in much of the eastern half of the U.S. Thanks to a global economy, such introductions seem likely to continue, keeping researchers busy for years to come.

Getting Involved

The vast majority of deaths due to mosquito-borne diseases happen outside of North America, but it's still possible for folks here to help. Many reputable, high-quality charities are leading the fight against malaria and other tropical diseases. Current malaria-prevention strategies involve distributing insecticide-infused bed nets, providing anti-malarial drugs, creating new and more effective repellents and continuing work on the Holy Grail of malaria research: an effective, cheap vaccine that could be used to eradicate the disease altogether.

The following charities are all highly rated:

Against Malaria: www.againstmalaria.com

UNICEF America: www.unicefusa.org

Oxfam International: www.oxfam.org

The Carnegie Institution for Science: http://carnegiescience.edu/ support/giving/research

CDC Foundation: www.cdcfoundation.org

Neglected Tropical Diseases: www.who. int/neglected_diseases/ diseases/en/

An *Anopheles* mosquito

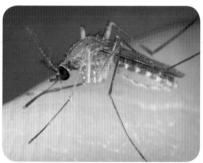
A *Culex* mosquito

Identifying Aedes, Anopheles and Culex Mosquitoes

Entomology is a complicated science, and it takes a considerable amount of knowledge (and magnification!) to identify individual mosquito species. Thankfully, the three primary groups (genera) of disease-carrying mosquitoes have some characteristic traits that can help amateurs identify them, or at least make an educated guess or two.

Of course, finding a potential disease-carrying mosquito doesn't necessarily mean that the mosquito is infected with a disease, and whichever type of mosquito you find, you'll probably still want to squash it.

Still, identifying species by genus is worthwhile, as it gives you firsthand knowledge of just how common such potential disease-carrying species are. Plus, when you see how many mosquitoes there really are, it makes it easier to understand why tropical diseases are such a persistent problem in much of the world.

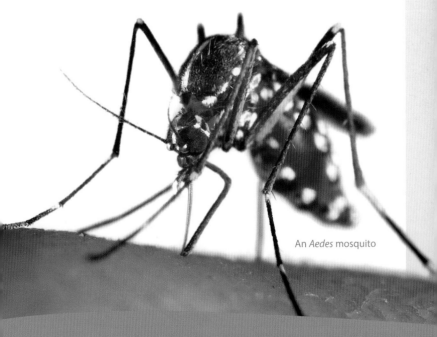

An *Aedes* mosquito

How to Identify Potential Disease-Carrying Mosquitoes

Identifying Eggs

Identification of individual mosquito species can get confusing. Thankfully, there are some general rules of thumb that can help you differentiate between the three general types of mosquitoes that can carry diseases in North America.

Anopheles mosquitoes lay their eggs singly[161] on the surface of the water. These eggs have a little bump on each side—these "air floats" are makeshift pontoons that help the eggs stay on the surface. *Anopheles* eggs also often occur in star-shaped patterns thanks to the surface tension of water[162].

Aedes lay their eggs singly, too, but the eggs lack the noticeable pontoon-like floats. They often lay their eggs above the waterline of a container[164] or in damp soil that will soon be flooded[163].

Culex mosquitoes lay their eggs on the water and actually glue them together, forming an egg raft. One other genus (*Culiseta*) also creates egg rafts, but its eggs tend to be larger.

Rule of Thumb: if you've spotted single eggs, take a closer look. If you've noted the "air floats," then you've found an *Anopheles* mosquito. If you spotted an egg raft, it's a good bet you've spotted a member of the *Culex* genus. Of the three, it's hardest to identify *Aedes* eggs.

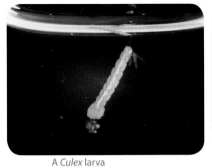

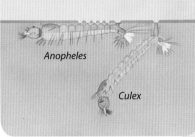

A *Culex* larva

Resting postures of
Anopheles and *Culex* larvae

Identifying Larvae

All mosquito larvae are aquatic, but they have different mechanisms for breathing, and this makes it fairly easy to tell *Aedes* and *Culex* larvae apart from the malaria-carrying *Anopheles*.

Aedes and *Culex* mosquitoes breathe via a siphon—essentially a snorkel—and hang from the surface of the water. *Culex* mosquitoes[165] have a longer siphon than those of *Aedes*.

Anopheles mosquitoes, however, are different. Instead of hanging down from the water, they are found parallel to the water, immediately below its surface. This enables them to breathe via tubes (spiracles) in their abdomen.

Identifying Adults

Anopheles mosquitoes differ from *Aedes* and *Culex* mosquitoes in an important respect: when they are resting, they have an entirely different body posture.

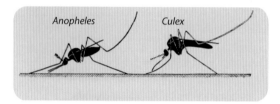

Anopheles   Culex

Brazil, Indonesia, Malaysia and Thailand have the greatest number of mosquito species present, with 447, 439, 415 and 379 species, respectively[167].

When resting, *Anopheles* mosquitoes have a steeply angled body posture relative to the surface. *Aedes* and *Culex* don't, so if you find a mosquito resting at an angle, it's probably an *Anopheles*.

So how do you determine between *Aedes* and *Culex* mosquitoes? That's a lot harder, but here's a tip: generally speaking, *Aedes* mosquitoes have pointed[166] abdomens; *Culex* mosquitoes have blunted abdomens.

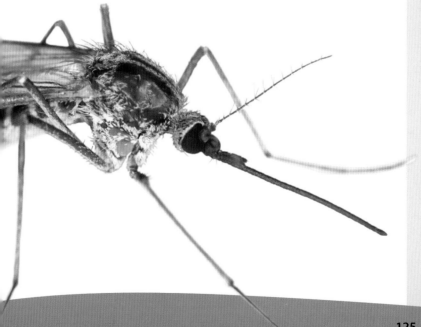

Table of Repellents/Insecticides

Name	Natural	Synthetic	Repellent	I want to prevent mosquitoes from biting me
DEET	No	Yes	Yes	Yes
Picaridin	No	Yes*	Yes	Yes
Permethrin (only use on clothing)	No	Yes*	Yes, but intended for use on clothing, not skin**	Yes
IR3535	Yes*	No	Yes	Yes
Oil of lemon eucalyptus/ PMD	Yes*	No	Yes	Yes

* Picaridin and Permethrin are inspired/based on naturally occurring chemical compounds. PMD is the synthesized version of the natural compound in oil of lemon eucalyptus. IR3535 is a synthetic version of a naturally occurring amino acid. If the man-made compound is the same as the one found in nature, I'm considering it natural.

I want to kill mosquitoes in my yard	Example brands/ products	Recommended by CDC as a topical repellent	Concentration strength
No	Repel, Off!, Coleman	Yes	Varies widely, from a few % to up to nearly 100 percent
No	Avon Skin So Soft Bug Guard Plus Picaridin, Natrapel, Sawyer, Off! FamilyCare	Yes	Varies from 5 percent up to 20 percent
Yes**	For clothing: Repel permethrin, Sawyer permethrin As an insecticide: Raid!	Permethrin is not a topical repellent, but it is recommended for clothing/gear	Varies depending on use; be sure to purchase the correct product for your use. Products used as insecticides often contain much higher concentrations
No	Many Avon products, Coleman, Sawyer	Yes	Varies from 7.5 percent up to 20 percent
No	Cutter, Citrepel, Off! Botanicals	Yes	Varies from 30 percent to 40 percent

** Permethrin has some repellent effect, but it also kills bugs on contact.
 Important note: Permethrin products for use on clothing are sold as sprays,
 but permethrin-based insecticides that are not intended for use on clothing
 are sold as sprays as well. Be sure that you're using the correct product.

Recommended Resources/Reading

American Mosquito Control Association
www.mosquito.org

Centers for Disease Control and Prevention
www.cdc.gov

Darsie, Jr., Richard F. and Ward, Ronald A.
Identification and Geographical Distribution of the Mosquitoes of North America, North of Mexico, 2004

Environmental Protection Agency Insect Repellent Search Tool
http://cfpub.epa.gov/oppref/insect/#searchform

Environmental Protection Agency: Pesticides
www.epa.gov/pesticides/index.htm

Extoxnet: Extension Toxicology Network
http://pmep.cce.cornell.edu/profiles/extoxnet/

Florida Medical Entomology Laboratory
http://fmel.ifas.ufl.edu

Malaria Journal
www.malariajournal.com

Mayo Clinic, Mosquito Bites: Prevention
www.mayoclinic.com/health/mosquito-bites/DS01075/
DSECTION=prevention

Mosquito Catalog
www.mosquitocatalog.org

PLOS ONE (scientific journal)
www.malariajournal.com

PubMed (database of peer-reviewed literature)
www.ncbi.nlm.nih.gov/pubmed

Walter Reed Biosystematics Unit
www.wrbu.org

Bibliography

1 Rey, Jorge. Florida Medical Entomology Laboratory, "The Mosquito." (Publication #ENY-727) http://edis.ifas.ufl.edu/in652

2 Commonwealth Science and Industrial Research Organization, "Insects and their Allies. Diptera - flies, mosquitoes." http://www.ento.csiro.au/education/insects/diptera.html

3 Gaffigan, Thomas, et al. Walter Reed Biosystematics Unit, "Systematic Catalog of Culicidae." http://www.mosquitocatalog.org/default.aspx?pgID=2

4 Foley, D. H., et al. "Insight into global mosquito biogeography from country species records." *Journal of Medical Entomology*. July 2007, Volume 44, Number 4, 554–567. http://www.ncbi.nlm.nih.gov/pubmed/17695008

5 Gaffigan, Thomas, et al. Walter Reed Biosystematics Unit, "Systematic Catalog of Culicidae." http://www.mosquitocatalog.org/default.aspx?pgID=2

6 Gaffigan, Thomas, et al. Walter Reed Biosystematics Unit, "Systematic Catalog of Culicidae." *Culicella minnesotae* (http://www.mosquitocatalog.org/taxon_descr.aspx?ID=16746) and *Culiseta alaskaensis* (http://www.mosquitocatalog.org/taxon_descr.aspx?ID=15331)

7 Walter Reed Biosystematics Unit, "Medically Important Mosquitoes." http://www.wrbu.org/northcom_MQ.html

8 Ibid.

9 Ibid.

10 Merriam-Webster, "*Anopheles*." http://www.merriam-webster.com/dictionary/culex

11 Merriam-Webster, "*Aedes*." http://www.merriam-webster.com/dictionary/culex

12 Merriam-Webster, "*Culex*." http://www.merriam-webster.com/dictionary/culex

13 Fang, Janet. "Ecology: A world without mosquitoes." *Nature*. 2010, 432–434. http://www.nature.com/news/2010/100721/full/466432a.html

14 Gaffigan, Thomas, et al. Walter Reed Biosystematics Unit, "Systematic Catalog of Culicidae." http://www.mosquitocatalog.org/default.aspx?pgID=2

15 Gjullin, C. M., et al. "Studies on *Aedes vexans* (Meig.) and *Aedes sticticus* (Meig.), Flood-water Mosquitoes, in the Lower Columbia River Valley." *Annals of the Entomological Society of America*. June 1950, Volume 43, Number 2, 262–275. http://lrsbeta.afpmb.org/smb/192.168.1.22/pdfs/Mosquito_Catalog/050600-0.PDF

16 Gaffigan, Thomas, et al. Walter Reed Biosystematics Unit, "Systematic Catalog of Culicidae." *Culex pipiens* (http://www.mosquitocatalog.org/taxon_descr.aspx?ID=18280)

17 American Mosquito Control Association, "Frequently Asked Questions." http://www.mosquito.org/faq

18 Gjullin, C. M., et al. "Studies on *Aedes vexans* (Meig.) and *Aedes sticticus* (Meig.), Flood-water Mosquitoes, in the Lower Columbia River Valley." *Annals of the Entomological Society of America*. June 1950, Volume 43, Number 2, 262–275.

19 Barr, A. R., et al. "The mosquito fauna of pitcher plants in Singapore." *Singapore Medical Journal*. 1963.

20 Goiny, H., et al., "The eggs of *Aedes* (skusea) *pembaensis* Theobold discovered on crabs." *East African Medical Journal*. 1957, Volume 34, 1–2.

21 Vezzani, D., "Review: Artificial container-breeding mosquitoes and cemeteries: a perfect match." *Tropical Medicine & International Health*. 2007, Volume 12, 299–313. doi: 10.1111/j.1365-3156.2006.01781.x

22 Ibid.

23 Mattingly, P. F. "Mosquito Eggs VI." *Mosquito Systematics Newsletter*. 1970, Number. 1, 17–22. www.mosquitocatalog.org/files/pdfs/MS02N01P017.pdf

24 Ibid.

25 Gjullin, C. M., et al. "Studies on *Aedes vexans* (Meig.) and *Aedes sticticus* (Meig.), Flood-water Mosquitoes, in the Lower Columbia River Valley." *Annals of the Entomological Society of America*. June 1950, Volume 43, Number 2, 262–275.

26 Lounibos, L. P., et al. *Mosquito Maternity: Egg Brooding in the Life Cycle of Trichoprosopon digitatum* (in *The Evolution of Insect Life Cycles*). 1986. http://link.springer.com/chapter/10.1007/978-1-4613-8666-7_11

27 De Meillon, B., et al. "The duration of egg, larval and pupal stages of *Culex pipiens fatigans* in Rangoon, Burma." *Bull World Health Organ*. 1967, Volume 36, Number 1, 7–14. http://www.ncbi.nlm.nih.gov/pubmed/4227199

28 Jackson, N. "Observations on the feeding habits of a predaceous mosquito larva, *Culex* (Lutzia) *Tigripes* Grandpre and Charmoy (Diptera)." *Proceedings of the Royal Entomological Society of London*. Series A, General Entomology, 1953, Volume 28, 153–159. doi: 10.1111/j.1365-3032.1953.tb00645.x

29 Goddard, Jerome. Bureau of General Environmental Services, Mississippi Department of Health, "Setting Up a Mosquito Control Program." Last modified June 2003. http://msdh.ms.gov/msdhsite/_static/resources/800.pdf

30 Centers for Disease Control and Prevention, "*Anopheles* Mosquitoes." Last modified November 09, 2012. http://www.cdc.gov/malaria/about/biology/mosquitoes/

31 Carpenter, Stanley J., and Walter J. Lacasse. *Mosquitoes of North America, North of Mexico.* Berkeley: University of California Press, 1955. http://www.mosquitocatalog.org/files/pdfs/016800-0.pdf

32 "South East Asia Mosquito Project." *Proceedings of the Entomological Society of Washington.* 1951, Number 1, 54. http://lrsbeta.afpmb.org/smb/192.168.1.22/pdfs/Mosquito_Catalog/074300-9-1.pdf

33 Zhou, Guoli, and Roger Miesfeld. "Differential utilization of blood meal amino acids in mosquitoes." *Open Access Insect Physiology.* 2009, 1–12. http://dx.doi.org/10.2147/OAIP.S7160

34 Greenberg, J. "Some nutritional requirements of adult mosquitoes (*Aedes aegypti*) for oviposition." *The Journal of Nutrition.* 1951, Volume 43, Issue 1, 27–35.

35 "Manual for Mosquito Rearing and Experimental Techniques." *AMCA Bulletin No. 5.* 1970. http://lrsbeta.afpmb.org/smb/192.168.1.22/pdfs/Mosquito_Catalog/048499-0.PDF

36 Greenberg, J. "Some nutritional requirements of adult mosquitoes (*Aedes aegypti*) for oviposition." *The Journal of Nutrition.* 1951, Volume 43, Issue 1, 27–35.

37 Howell, P. I., et al. "Male mating biology." *Malaria Journal.* November, 16, 2009, Volume 8, Supplement 2. doi: 10.1186/1475-2875-8-S2-S8.

38 Butail, Sachit, et al. "The Dance of Male *Anopheles gambiae* in Wild Mating Swarms." *Journal of Medical Entomology.* 2013, Volume 50, Number 3, 552–559. http://www.bioone.org/doi/abs/10.1603/ME12251

39 Lang, Susan. *Cornell Chronicle,* "Mosquitoes create harmonic love song before mating, a Cornell study finds." Last modified January 08, 2009. http://www.news.cornell.edu/stories/2009/01/study-mosquitoes-beat-out-love-song-mating

40 Clements, Alan N. *The Biology of Mosquitoes: Sensory Reception and Behaviour.* Chapman and Hall, 1999.

41 Carpenter, Stanley J., and Walter J. Lacasse. *Mosquitoes of North America, North of Mexico.* Berkeley: University of California Press, 1955. http://www.mosquitocatalog.org/files/pdfs/016800-0.pdf

42 VanDyk, John. Iowa State University, "Mosquito Host-Seeking: a partial review." http://www.ent.iastate.edu/dept/research/vandyk/hostseek.html

43 Department of Entomology, Rutgers University, "FAQ's on Mosquitoes." Last modified July 01, 2011. http://www-rci.rutgers.edu/~insects/mosfaq.htm

44 Brett, G. A. "On the Relative Attractiveness to *Aedes aegypti* of Certain Coloured Cloths." *Transactions of the Royal Society of Tropical Medicine and Hygiene.* 1938, 113–124.

45 VanDyk, John. Iowa State University, "Mosquito Host-Seeking: a partial review." http://www.ent.iastate.edu/dept/research/vandyk/hostseek.html

46 Wright, R. H., et al. "Some responses of flying *Aedes aegypti* to visual stimuli." *Canadian Entomology.* 1968, Volume 100, 504–513.

47 Jones, M. D. R., et al. "The Circadian Rhythm of Flight Activity of the Mosquito *Anopheles gambiae*: The Light-Response Rhythm." *The Journal of Experimental Biology.* 1972, 337–346. http://jeb.biologists.org/content/57/2/337.full.pdf%20html

48 Kelly, D.W. "Why are some people bitten more than others?" *Trends in Parasitology.* December 2001, Volume 17, Issue 12, 578–581. http://www.ncbi.nlm.nih.gov/pubmed/11756041

49 Cummins, B., et al. "A Spatial Model of Mosquito Host-Seeking Behavior." *PLOS Computational Biology.* 2012, Volume 8, Issue 5: e1002500. doi:10.1371/journal.pcbi.1002500.

50 Todar, Kenneth. *Todar's Online Textbook of Bacteriology.* "The Normal Bacterial Flora of Humans." http://www.textbookofbacteriology.net/normalflora_3.html

51 Verhulst, N. O., et al. "Composition of Human Skin Microbiota Affects Attractiveness to Malaria Mosquitoes." *PLoS ONE,* 2011, Volume 6, Issue 12: e28991. doi:10.1371/journal.pone.0028991

52 Todar, Kenneth. *Todar's Online Textbook of Bacteriology.* "The Normal Bacterial Flora of Humans." http://www.textbookofbacteriology.net/normalflora_3.html

53 Lefèvre, T., et al. "Beer Consumption Increases Human Attractiveness to Malaria Mosquitoes." *PLoS ONE,* 2010, Volume 5, Issue 3: e9546. doi:10.1371/journal.pone.0009546

54 Logan, James G., et al. "Arm-in-cage testing of natural human-derived mosquito repellents." *Malaria Journal.* 2010, Volume 9, Issue 239. Published online August 20, 2010. doi: 10.1186/1475-2875-9-239

55 Wang, Shirley. "Finding Smells That Repel." *The Wall Street Journal.* September 1, 2009. http://online.wsj.com/article/SB10001424052970204660604574378933761528214.html

56 Royal Chemical Society, *Chemical News*, "Keeping mosquitoes at bay." Last modified November 06, 2006. http://www.rsc.org/chemistryworld/News/2006/November/06110601.asp

57 Cummins, B., et al. "A Spatial Model of Mosquito Host-Seeking Behavior." *PLOS Computational Biology.* 2012, Volume 8, Issue 5: e1002500. doi:10.1371/journal.pcbi.1002500

58 Ritchie, S. A., et al. "Wind-blown mosquitoes and introduction of Japanese encephalitis into Australia." *Emerging Infectious Diseases.* 2001, Volume 7, 900–903.

59 Cummins, B., et al. "A Spatial Model of Mosquito Host-Seeking Behavior." *PLOS Computational Biology.* 2012, Volume 8, Issue 5: e1002500. doi:10.1371/journal.pcbi.1002500

60 Lindsay, S., et al. "Effect of pregnancy on exposure to malaria mosquitoes." *Lancet.* June 3, 2000, Volume 355, Issue 9219, 1972.

61 Meyer, John. NC State University, "Entomology. External Anatomy: Mouthparts." Last modified May 1, 2013. http://www.cals.ncsu.edu/course/ent425/

62 Ramasubramanian, M. K., et al. "Mechanics of a mosquito bite with applications to microneedle design." *Bioinspiration and Biomimetics.* December 2008, Volume 3, Issue 4, 046001. doi: 10.1088/1748-3182/3/4/046001.

63 Calvo, E., et al. "The salivary gland transcriptome of the eastern tree hole mosquito, *Ochlerotatus triseriatus.*" *Journal of Medical Entomology.* May 2010, Issue 47, Volume 3, 376–386.

64 NASA, "Warp Drive, When?" http://www.nasa.gov/centers/glenn/technology/warp/warpstat.html

65 Kim, B. H., et al. "Experimental analysis of the blood-sucking mechanism of female mosquitoes." *Journal of Experimental Biology.* April 1, 2011, Volume 214 (part 7), 1163–1169. doi: 10.1242/jeb.048793

66 Berenbaum, May. *Buzzwords.* "Mosquito myth exploded?" *American Entomologist.* http://www.entsoc.org/PDF/Pubs/Periodicals/AE/AE-2009/Spring/Buzz.pdf

67 Gwadz, Robert W. "Regulation of blood meal size in the mosquito," *Journal of Insect Physiology.* November 1969, Volume 15, Issue 11, 2039–2042, http://dx.doi.org/10.1016/0022-1910(69)90071-7 (http://www.sciencedirect.com/science/article/pii/0022191069900717)

68 Burns, Robert, and Dr. Jimmy Olson. "Many mosquito controls only hammer homeowner's pocketbook." *Agrilife Extension, Texas A&M.* http://texashelp.tamu.edu/004-natural/pdfs/mosquito-control-options-and-gimmicks.pdf

69 Silver, John B. *Mosquito Ecology: Field Sampling Methods.* New York: Springer, 2008. http://www.springer.com/life sciences/entomology/book/978-1-4020-6665-8

70 Reisen, W. K., et al. "Effects of sampling design on the estimation of adult mosquito abundance." *Journal of the American Mosquito Control Association.* June 1999, Volume 15, Issue 2, 105–114.

71 Coachella Valley Economic Partnership, "Demographics." Last modified 2013. http://cvep.com/demographics.htm

72 Read, N. R. "Public perception of mosquito annoyance measured by a survey and simultaneous mosquito sampling." *Journal of the American Mosquito Control Association*. March 1994, Volume 10, Issue 1, 79–87.

73 American Red Cross, "Blood Facts and Statistics." http://www.redcrossblood.org/learn-about-blood/blood-facts-and-statistics

74 Bowman, Dwight, and Jay Georgi. *Georgis' Parasitology for Veterinarians*. Elsevier Health Sciences, 2009. http://books.google.com/books?id=g_tBWVBevM0C&dq=cattle exsanguination mosquitoes&source=gbs_navlinks_s

75 Abbitt, B., et al. "Fatal exsanguination of cattle attributed to an attack of salt marsh mosquitoes (*Aedes sollicitans*)." *Journal of the American Veterinary Medicine Association*. December 15, 1981, Volume 179, Issue 12, 397–400.

76 Brune, Jeff. U.S. Department of Interior, Bureau of Land Management Alaska, "Surviving the Arctic Tundra: A Look at Cold-Weather Adaptations." http://www.blm.gov/ak/st/en/res/education/akcold_desert/akcolddesert_posterback.print.html

77 Addison, David, and Scott Ritchie. "Cattle Fatalities From Prolonged Exposure to *Aedes taeniorhynchus* in Southwest Florida." *Florida Scientist*. 1993, 65–69. http://archive.org/stream/floridascie5657199394flor

78 National Institutes of Health, "Life Cycle of the Malaria Parasite." Last modified April 03, 2012. http://www.niaid.nih.gov/topics/malaria/pages/lifecycle.aspx

79 Muriu, S. M., et al. "Host choice and multiple blood feeding behaviour of malaria vectors and other anophelines in Mwea rice scheme, Kenya." *Malaria Journal*. February 29, 2008, Volume 7, Issue 43. doi: 10.1186/1475-2875-7-43

80 Entomological Society of America, "West Nile virus Passes from Female to Eggs." Last modified March 05, 2013. http://www.entsoc.org/press-releases/west-nile-virus-passes-female-eggs

81 Kirk, Catherine, et al. "Twin Study of Adolescent Genetic Susceptibility to Mosquito Bites Using Ordinal and Comparative Rating Data." *Genetic Epidemiology*. 2000, 178–190. http://genepi.qimr.edu.au/contents/p/staff/CV270.pdf

82 Shirai, Y., et al. "Landing preference of *Aedes albopictus* (Diptera: Culicidae) on human skin among ABO blood groups, secretors or nonsecretors, and ABH antigens." *Journal of Medical Entomology*. July 2004, Volume 41, Issue 4, 796–799.

83 Carter, R., et al. "Evolutionary and historical aspects of the burden of malaria." *Clinical Microbiology Review.* October 2002, Volume 15, Issue 4, 564–594.

84 Centers for Disease Control and Prevention, "About Malaria." Last modified February 02, 2010. http://www.cdc.gov/malaria/about/disease.html

85 Ibid.

86 Ibid.

87 Carter, R., et al. "Evolutionary and historical aspects of the burden of malaria." *Clinical Microbiology Review.* October 2002, Volume 15, Issue 4, 564–594.

88 Centers for Disease Control and Prevention, "About Malaria, Biology." Last modified February 02, 2010. http://www.cdc.gov/malaria/about/biology/

89 Wiser, Mark. Tulane University, "Malaria." Last modified November 15, 2011. http://www.tulane.edu/~wiser/protozoology/notes/malaria.html

90 Whiteford, John Noble. "Malaria Is a Likely Killer in King Tut's Post-Mortem." *New York Times.* February 16, 2010.

91 Carter, R., et al. "Evolutionary and historical aspects of the burden of malaria." *Clinical Microbiology Review.* October 2002, Volume 15, Issue 4, 564–594.

92 Ibid.

93 Centers for Disease Control and Prevention, "Mortality Statistics, 1900–1904." http://www.cdc.gov/nchs/data/vsushistorical/mortstatsh_1900-1904.pdf

94 World Health Organization, "10 Facts on Malaria." Last modified March 2013. http://www.who.int/features/factfiles/malaria/en/

95 Murray, C. J., et al. "Global malaria mortality between 1980 and 2010: a systematic analysis." *Lancet.* February 4, 2012, Volume 379, Issue 9814, 413–431. doi: 10.1016/S0140-6736(12)60034-8

96 Silberner, Joanne. "WHO Backs Use of DDT Against Malaria." *National Public Radio.* September 15, 2006. http://www.npr.org/templates/story/story.php?storyId=6083944

97 Carter, R., et al., "Evolutionary and historical aspects of the burden of malaria." *Clinical Microbiology Review.* October 2002, Volume 15, Issue 4, 564–594.

98 Andrianaivolambo, L., et al. "Anthropophilic mosquitoes and malaria transmission in the eastern foothills of the central highlands of Madagascar." *Acta Tropica.* December 2010, Volume 116, Issue 3, 240–245. doi: 10.1016/j.actatropica.2010.08.017

99 Poinar, George, Jr. "*Plasmodium dominicana* n. sp. (Plasmodiidae: Haemospororida) from Tertiary Dominican amber." *Systematic Parasitology.* 2005, 47–52. http://link.springer.com/article/10.1007/s11230-004-6354-6

100 Carter, R., et al. "Evolutionary and historical aspects of the burden of malaria." *Clinical Microbiology Review.* October 2002, Volume 15, Issue 4, 564–594.

101 Quinn, C. T., et al. "Survival of children with sickle cell disease." *Blood.* June 1, 2004, Volume 103, Issue 11, 4023-4027.

102 Carter, R., et al. "Evolutionary and historical aspects of the burden of malaria." *Clinical Microbiology Review.* October 2002, Volume 15, Issue 4, 564–594.

103 Centers for Disease Control and Prevention, "About Malaria, History." Last modified February 02, 2010. http://www.cdc.gov/malaria/about/history/

104 Murray, C. J., et al. "Global malaria mortality between 1980 and 2010: a systematic analysis." *Lancet.* February 4, 2012, Volume 379, Issue 9814, 413–431. doi: 10.1016/S0140-6736(12)60034-8

105 National Institutes of Health, "Investigational malaria vaccine found safe and protective." Last modified August 08, 2013. http://www.nih.gov/news/health/aug2013/niaid-08.htm

106 Ferguson. H. M., et al. "Why is the effect of malaria parasites on mosquito survival still unresolved?" *Trends in Parasitology.* June 2002, Volume 18, Issue 6, 256–261.

107 Centers for Disease Control and Prevention, "Yellow Fever: Symptoms and Treatment." Last modified December 13, 2011. http://www.cdc.gov/yellowfever/symptoms/index.html

108 Centers for Disease Control and Prevention, "Dengue: Frequently Asked Questions." Last modified September 27, 2012. http://www.cdc.gov/dengue/fAQFacts/index.html

109 Berman, Jessica. Voice of America, "Dengue Fever Vaccine Trials Clear First Hurdle." Last modified January 24, 2013. http://www.voanews.com/content/dengue-fever-vaccine/1590524.html

110 Centers for Disease Control and Prevention, "West Nile virus disease cases and deaths reported to CDC by year and clinical presentation, 1999–2012." http://www.cdc.gov/westnile/resources/pdfs/cummulative/99_2012_CasesAndDeathsClinicalPresentationHumanCases.pdf

111 Murray, K. O. "West Nile virus and its emergence in the United States of America. (Review)." *Veterinary Research*. November–December, 2010, Volume 41, Issue 6, 67.

112 Centers for Disease Control and Prevention, "Fact Sheet: Arboviral Encephalitis." Last modified February 9, 2009. http://www.cdc.gov/ncidod/dvbid/arbor/arbofact.htm

113 The Heartworm Society, "What Is Heartworm Disease?" http://www.heartwormsociety.org/pet-owner-resources/heartworm.html

114 The U.S. Food and Drug Administration, "Keep The Worms Out Of Your Pet's Heart! The Facts About Heartworm Disease." Last modified June 06, 2013. http://www.fda.gov/animalveterinary/resourcesforyou/animalhealthliteracy/ucm188470.htm

115 Brown, A. W. A. "Studies on the Responses of the Female *Aëdes* Mosquito. Part VI.—The Attractiveness of coloured Cloths to Canadian Species." *Bulletin of Entomological Research*. 1954, Volume 45, 67–78. doi:10.1017/S0007485300026808

116 Dickens, Joseph C., et al. "Mini review: Mode of action of mosquito repellents." *Pesticide Biochemistry and Physiology*, Volume 106, Issue 3, July 2013, 149–155, http://dx.doi.org/10.1016/j.pestbp.2013.02.006. (http://www.sciencedirect.com/science/article/pii/S0048357513000308

117 Ibid.

118 The Mayo Clinic, "Mosquito bites: prevention." Last modified October 24, 2012. http://www.mayoclinic.com/health/mosquito-bites/DS01075/DSECTION=prevention

119 Antwi, F. B., et al. "Risk assessments for the insect repellents DEET and picaridin." *Regulatory Toxicology and Pharmacology*. June 2008, Volume 51, Issue 1, 31–36. doi: 10.1016/j.yrtph.2008.03.002. Epub March 2008.

120 Environmental Protection Agency, "Reregistration Eligibility Decision (RED) DEET." Last modified September 1998. http://www.epa.gov/oppsrrd1/REDs/0002red.pdf

121 National Pesticide Information Center, "DEET Fact Sheet." Last modified September 1998. http://npic.orst.edu/factsheets/DEETtech.pdf

122 Maia, M. F., et al. "Plant-based insect repellents: a review of their efficacy, development and testing." *Malaria Journal*. March 2011, Volume 15, Issue 10, Supplement 1, Section 11. doi: 10.1186/1475-2875-10-S1-S11

123 Zhu, J. J., et al. "Repellency of a Wax-Based Catnip-Oil Formulation against Stable Flies." *Journal of Agricultural and Food Chemistry*. November 8, 2010.

124 Environmental Protection Agency, "Pesticides: Regulating Pesticides, Pyrethroids and Pyrethrins." Last modified April 2013. http://www.epa.gov/oppsrrd1/reevaluation/pyrethroids-pyrethrins.html

125 Cornell University, "Extension Toxicology Network: Permethrin." Last modified September 1993. http://pmep.cce.cornell.edu/profiles/extoxnet/metiram-propoxur/permethrin-ext.html

126 Óscar López, ed. *Green Trends in Insect Control* (Chapter 3 Pyrethrins and Pyrethroid Insecticides, Schleier III, Jerome, and Robert Peterson). London: Royal Society of Chemistry, 2011. http://entomology.montana.edu/People/RKDPeterson/Schleier III and Peterson 2011 (Pyrethrins and Pyrethroids Chapter).pdf

127 Cornell University, "Extension Toxicology Network: Permethrin." Last modified September 1993. http://pmep.cce.cornell.edu/profiles/extoxnet/metiram-propoxur/permethrin-ext.html

128 "Permethrin-Treated Army Combat Uniforms." *Stand-To!* October 10, 2012. http://www.army.mil/standto/archive/issue.php?issue=2012-10-10

129 Nasci, Roger, et al. Centers for Disease Control and Prevention, "Protection against Mosquitoes, Ticks, & Other Insects & Arthropods." Last modified August 08, 2013. http://wwwnc.cdc.gov/travel/yellow-book/2014/chapter-2-the-pre-travel-consultation/protection-against-mosquitoes-ticks-and-other-insects-and-arthropods

130 Schreck, C. E., et al. "The effectiveness of permethrin and deet, alone or in combination, for protection against *Aedes taeniorhynchus.*"*American Journal of Tropical Medicine and Hygiene.* July 1984, Volume 33, Issue 4, 725–730.

131 National Pesticide Information Center, "Permethrin Treated Clothing." Last modified February 28, 2013. http://npic.orst.edu/pest/mosquito/ptc.html

132 Óscar López, ed. *Green Trends in Insect Control* (Chapter 3, Pyrethrins and Pyrethroid Insecticides, Schleier III, Jerome, and Robert Peterson). London: Royal Society of Chemistry, 2011. http://entomology.montana.edu/People/RKDPeterson/Schleier III and Peterson 2011 (Pyrethrins and Pyrethroids Chapter).pdf

133 National Pesticide Information Center, "Picaridin General Fact Sheet." Last modified December 2009. http://npic.orst.edu/factsheets/PicaridinGen.html

134 Nasci, Roger, et al. Centers for Disease Control and Prevention, "Protection against Mosquitoes, Ticks, & Other Insects & Arthropods." Last modified August 08, 2013. http://wwwnc.cdc.gov/travel/yellow-book/2014/chapter-2-the-pre-travel-consultation/protection-against-mosquitoes-ticks-and-other-insects-and-arthropods

135 The Mayo Clinic, "Mosquito bites: prevention." Last modified October 24, 2012. http://www.mayoclinic.com/health/mosquito-bites/DS01075/DSECTION=prevention

136 Moreton Bay (Australia) Regional Council, "Koalas: Fact Sheet." http://www.moretonbay.qld.gov.au/uploadedFiles/moretonbay/environment/fauna/KoalasFactSheet.pdf

137 Maia, M. F., et al. "Plant-based insect repellents: a review of their efficacy, development and testing." *Malaria Journal*. March 2011, Volume 15, Issue 10, Supplement 1, Section 11. doi: 10.1186/1475-2875-10-S1-S11

138 Environmental Protection Agency, "p-Menthane-3,8-diol (011550) Fact Sheet." http://www.epa.gov/opp00001/chem_search/reg_actions/registration/fs_PC-011550_01-Apr-00.pdf

139 Nasci, Roger, et al. Centers for Disease Control and Prevention, "Protection against Mosquitoes, Ticks, & Other Insects & Arthropods." Last modified August 08, 2013. http://wwwnc.cdc.gov/travel/yellow-book/2014/chapter-2-the-pre-travel-consultation/protection-against-mosquitoes-ticks-and-other-insects-and-arthropods

140 Environmental Working Group, "EWG's Guide to Bug Repellents: Repellent Chemicals." http://www.ewg.org/research/ewgs-guide-bug-repellents/repellent-chemicals

141 Environmental Protection Agency, "Completed IR3535 Insect Repellent Efficacy Studies." Last modified March 02, 2007. http://www.epa.gov/hsrb/files/meeting-materials/apr-18-20-2007-public-meeting/DRAFTFinalHSRBReportOnEMD-004.1-4.2CompletedStudies.pdf

142 Centers for Disease Control and Prevention, "FAQ: Insect Repellent Use & Safety." Last modified June 07, 2013. http://www.cdc.gov/westnile/faq/repellent.html

143 Fradin, M.S., et al. "Comparative efficacy of insect repellents against mosquito bites." *New England Journal of Medicine*. July 2002, Volume 4, Issue 347, 13–18.

144 Klocke, James, et al. "1,8-Cineole (Eucalyptol), a mosquito feeding and ovipositional repellent from volatile oil of *Hemizonia fitchii* (Asteraceae)." *Journal of Chemical Ecology*. 1987, Volume 12, 2131–2141. http://link.springer.com/article/10.1007/BF01012562

145 Rutledge, Roxanne, and Jonathan Day. Florida Medical Entomology Laboratory, "Mosquito Information Website (Mosquito Repellents)." Last modified September 2002. http://mosquito.ifas.ufl.edu/Mosquito_Repellents.htm

146 American Mosquito Control Association, "Frequently Asked Questions." http://www.mosquito.org/faq

147 Ibid.

148 Foster, W. A., et al. "Tests of ultrasonic emissions on mosquito attraction to hosts in a flight chamber." *Journal of the American Mosquito Control Association*. June 1985, Volume 1, Issue 2, 199–202.

149 American Mosquito Control Association, "Frequently Asked Questions." http://www.mosquito.org/faq

150 Extension Toxicology Network, *"Bacillus thuringiensis."* Last modified May 1994. http://pmep.cce.cornell.edu/profiles/extoxnet/24d-captan/bt-ext.html

151 Lindsay, L. R. "Evaluation of the efficacy of 3% citronella candles and 5% citronella incense for protection against field populations of *Aedes* mosquitoes." *Journal of the American Mosquito Control Association*. June 1996, Volume 12, Issue 2, Part 1, 293–294.

152 Markowitz, Mike. Defense Media Network, "Men Against Mosquitoes: Malaria in War." Last modified February 02, 2013. http://www.defensemedianetwork.com/stories/men-against-mosquitoes-malaria-in-war/

153 U.S. Army Chemical Corps, "Summary of Major Events and Problems." Last modified 1960. https://www.osti.gov/opennet/servlets/purl/16006843-5BAfk6/16006843.pdf

154 Rose, William. U.S. Army Chemical Corps, "An Evaluation of Entomological Warfare as a Potential Danger to the United States and European NATO Nations." Last modified March 1981. http://www.thesmokinggun.com/documents/crime/attack-killer-mosquitoes-0

155 Feinberg, Ashley. *Gizmodo*, "The Entire Premise of Jurassic Park Is Wrong Because of Mosquitoes." Last modified July 07, 2013. http://gizmodo.com/the-entire-premise-of-jurassic-park-is-wrong-because-of-961667776

156 Berenbaum, May. *Buzzwords*. "Mosquito myth exploded?" *American Entomologist*. http://www.entsoc.org/PDF/Pubs/Periodicals/AE/AE-2009/Spring/Buzz.pdf

157 Hill, Stephanie, and C. Roxanne Connelly. University of Florida Entomology Department, "Human Bot Fly (*Dermatobia hominis*)." Last modified July 2008. http://entnemdept.ufl.edu/creatures/misc/flies/human_bot_fly.htm

158 Ibid.

159 Reinert, J. F., et al. "Phylogeny and classification of Aedini (Diptera: Culicidae), based on morphological characters of all life stages." *Zoological Journal of the Linnean Society*, 2004, Volume 142, 289–368. doi: 10.1111/j.1096-3642.2004.00144.x

160 Turner, Stephanie Lynn, et al. "Ultra-prolonged activation of CO_2-sensing neurons disorients mosquitoes." *Nature*. 2011, 87–91. http://www.nature.com/nature/journal/v474/n7349/full/nature10081.html

161 New Mexico Department of Health, "Classification and Identification of Mosquitoes of New Mexico." http://nmhealth.org/ERD/HealthData/documents/a_CLASSIFICATIONANDID_000.pdf

162 Carpenter, Stanley J., and Walter J. Lacasse. *Mosquitoes of North America, North of Mexico*. Berkeley: University of California Press, 1955. http://www.mosquitocatalog.org/files/pdfs/016800-0.pdf

163 American Mosquito Control Association, "Life Cycle of the Malaria Parasite." http://www.mosquito.org/life-cycle

164 New Mexico Department of Health, "Classification and Identification of Mosquitoes of New Mexico." http://nmhealth.org/ERD/HealthData/documents/a_CLASSIFICATIONANDID_000.pdf

165 Department of Medical Entomology, University of Sydney and Westmead Hospital, Australia, "Mosquito Larvae Photos." http://medent.usyd.edu.au/photos/larvae_photographs.htm

166 Walter Reed Biosystematics Unit, "Keys Tutorial, Adult Female Mosquito: Abdomen, dorsal view." http://www.wrbu.org/tut/adult_tax_tut16.html

167 Foley, D. H., et al. "Insight into global mosquito biogeography from country species records." *Journal of Medical Entomology*, July 2007, Volume 44, Issue 4, 554–567.

Photo Credits

Many of the images in this book originate from the Centers for Disease Control and Prevention's Public Health Imagery Library (http://phil.cdc.gov/phil/home.asp), a fantastic resource for high-quality photos.

17, 21 Prof. Woodbridge Foster, Prof. Frank H. Collins, photos by James Gathany. **29** (*Anopheles* and *Aedes*) CDC, (*Culex*) Harry Weinburgh. **31** Harry Weinburgh. **43** Segrid McAllister/Janice Haney Carr. **53** (inset), **54** Janice Haney Carr, photos by Elizabeth Perez. **71** (left) Dr. Mae Melvin, (right) Steven Glenn, Laboratory & Consultation Division. **72** Steven Glenn, Laboratory & Consultation Division. **74** Prof. Frank Hadley Collins, Dir., Cntr. for Global Health and Infectious Diseases, Univ. of Notre Dame, James Gathany. **75** Sickle Cell Foundation of Georgia: Jackie George, Beverly Sinclair.
79 Prof. Frank Hadley Collins, Dir., Cntr. for Global Health and Infectious Diseases, Univ. of Notre Dame, James Gathany. **122** (*Anopheles* and *Aedes*) CDC, (*Culex*) Harry Weinburgh. **123** (left) Dr. Pratt, (comparison) CDC.

James Gathany, Univ. of Notre Dame: **20** (both), **26, 27, 30, 32, 50, 118** (left), **120** (left), **121, 143**

The public domain images on the following page were accessed from the National Institutes of Health from the History of Medicine. (http://www.nlm.nih.gov/hmd/ihm/): **112**

The images on the following pages are licensed according to the Creative Commons 3.0 Attribution License, available at http://creativecommons.org/licenses/by/3.0/us/

56 (mosquito proboscis), courtesy of Ben Knight-Gregson. **81** (heartworm micrograph), courtesy of Joel Mills. **91** (*Tancacetum cinerariifolium*), courtesy of Roger Culos

All other images are public domain or are from stock photography sources.

About the Author

Brett Ortler is an editor at Adventure Publications. While at Adventure, he has edited dozens of books, including many field guides and nature-themed books. In addition to this book, he's authored *The Firefly Book*. His own work appears widely, including in *Salon*, *The Good Men Project*, *The Nervous Breakdown*, *Living Ready* and in a number of other venues in print and online. He lives in the Twin Cities with his wife and their young son.